"For those psychonauts out there wanting to read only about psychedelics, take note. The discovery of LSD did not 'just happen.' It emerged scientifically from a long and methodical study of ergot alkaloids by Albert Hofmann, who was neither looking for, nor previously interested in, anything to do with psychedelic drugs. After the discovery of the psychoactive effects of LSD, Hofmann, of course, went on to dedicate his life to psychedelics. But apart from a brief and enlightening discussion in the closing chapter, this book is about the chemistry of ergot alkaloids—not psychedelics. Nevertheless, the importance of psychedelics grows, mushroom-like, from the pages as one reads the fascinating journey from ergot to lysergic acid to lysergic acid diethylamide that eventually took Hofmann into the history books. And that makes this seminal 1964 book of Hofmann's an absolute treasure for any serious psychonaut."

—Ben Sessa, MBBS (MD) BSc MRCPsych, Psychiatrist,
Psychedelic Researcher, and Author of *The Psychedelic Renaissance*

"This book is a completely fascinating and detailed technical account of the chemical, physiological, and personal investigations of Hofmann and others with a whole range of ergot-related compounds, several of which have powerful psychedelic effects, not only the star of this book, LSD."

—Rupert Sheldrake, PhD, Author of *Ways to Go Beyond and Why They Work*

"One doesn't have to be a chemist to appreciate the translation of this classic work. It would be hard to overstate the impact of Albert Hoffman's work with ergot alkaloids on modern medicine, on Western culture in general, and on the recent resurgence of interest in the therapeutic potential of psychedelics. It's remarkable that the chemist who first accidentally ingested LSD happened to be someone with the curiosity and courage to experiment further and the openness and wisdom to grasp the importance of his subjective experience."

—Michael Mithoefer, MD, MDMA Researcher

"Dr. Hofmann's work remains among the most lucid and influential in psychedelic science and chemistry. Researchers and enthusiasts from all backgrounds can learn from and enjoy his cogent insights, which helped lay the groundwork for the current psychedelic renaissance. *Ergot Alkaloids* provides a thought-provoking window into the evolution of psychedelic science for contemporary readers."

—Albert Garcia-Romeu, PhD, Johns Hopkins Center for
Psychedelic and Consciousness Research

"There is no doubt that Hofmann changed not only psychopharmacology but also neuroscience with his ergot chemistry. The current renaissance of psychedelic chemistry is homage to his breakthrough contributions, and those newly in the field will benefit from the knowledge and historical perspectives contained in this timely republishing of a classic text."

—David Nutt, PhD, Prof of Neuropsychopharmacology and Head of the Centre for Psychedelic Research Imperial College London

"*Ergot Alkaloids*, published 33 years before Sasha and Ann Shulgin's *TIHKAL*, describes the synthesis of various lysergamides and description of plants containing them as well as other psychedelics that are found in plants, such as mescaline in peyote, psilocybin in psilocybe mushrooms, and lysergic acid in ololiuqui seeds. This is a rare psychedelic classic now available to English-language readers."

—George Greer, MD, President, Heffter Research Institute

"Hofmann was one of the world's most renowned natural products chemists. His knowledge and expertise in the field of ergot alkaloids and their chemistry were unmatched during his long career at Sandoz. As we know, Hofmann was the discoverer of the potent psychedelic agent LSD (lysergic acid diethylamide). He briefly discusses that discovery, but this book is so much more. Starting with A. Stoll's 1918 isolation and identification of ergotamine, Hofmann provides encyclopedic coverage of the known alkaloids extracted from ergot fungus (*Claviceps*), including their structural elucidation and how they were extracted, whether from infected grains from different geographical regions, or from cultures of the fungus, and provides analytical data characterizing each alkaloid. Included also are examples of medical uses for many ergot alkaloids and their derivatives."

—David E. Nichols, Distinguished Professor Emeritus, Purdue University School of Pharmacy

"On the occasion of the publication of the English translation of Albert Hofmann's classic text *Ergot Alkaloids: Their History, Chemistry, and Therapeutic Uses*, it is timely to reflect on the extraordinary contributions this esteemed natural products chemist made to the evolution of modern pharmacology, medicine, and psychiatry. From Hofmann's serendipitous discovery of LSD in 1943 to his isolation of psilocybin from samples of hallucinogenic mushrooms provided to him by R. Gordon Wasson's field explorations in Oaxaca, Mexico, in the late 1950s, Hofmann made pivotal discoveries to our understanding of powerful consciousness altering drugs that in recent years have become recognized as potentially among the most valuable of our psychopharmacologic armamentarium. Long neglected and often feared for their range of profound mindaltering effects, Hofmann's remarkable discoveries are now beginning to attract the recognition and respect they have long deserved for their unique therapeutic range of action and innate potential to facilitate deep explorations of inner psychic terrain that are increasingly recognized as being of great value to our understanding of comparative religion, cultural anthropology, depth psychology, and neuroscience. Whereas efforts designed to utilize these powerful ergot derived psychedelic alkaloids in psychiatric treatments in the 1950s and 1960s proved difficult to sustain because of ingrained cultural resistance, renewed laboratory and clinical research activity in recent years has proved more successful at catalyzing the level of mainstream societal support that will allow its therapeutic potentials to be more fully explored. Albert Hofmann, who passed away in 2008 at the age of 102, will likely be remembered by history as one of the most accomplished scientist-philosophers of the 20th century, whose inestimable contributions to the betterment of humanity and the natural world will finally receive the full appreciation that he has long merited. This classic text of his laboratory explorations and discoveries will further catalyze well-earned attention to Hofmann's status as the preeminent natural products research chemist of his day, whose discoveries and contributions will continue to reverberate well into the future."

—Charles S. Grob, M.D., Professor of Psychiatry and Biobehavioral Sciences and Pediatrics, David Geffen School of Medicine at UCLA

ERGOT
ALKALOIDS

ERGOT ALKALOIDS

HISTORY, CHEMISTRY,
AND THERAPEUTIC USES

ALBERT HOFMANN

Translated by JITKA NYKODEMOVÁ

Foreword by AMANDA FEILDING
Afterword by WILLIAM LEONARD PICKARD

TRANSFORM PRESS

Co-published by
Transform Press | PO Box 11552, Berkeley, California 94712
Synergetic Press | 1 Blue Bird Court, Santa Fe, New Mexico 87508
& 24 Old Gloucester Street, London WCIN 3AL, England

The Library of Congress Cataloging-in-Publication data is available from the Library of Congress.

ISBN 9798987497708 (hardcover)
ISBN 9798987497715 (ebook)

Originally published as *Die Mutterkornalkaloide: Vom Mutterkorn zum LSD—Die Chemie der Mutterkornalkaloide* © 2000 Nachtschatten Verlag

Chemistry consultant: Paul Daley

Cover design: Amanda Müller
Book design: David Good Design
Typesetting: Jonathan Hahn
Managing Editor: Noelle Armstrong
Production Editor: Allison Felus

Printed in the United States of America

CONTENTS

FOREWORD

The story of how Albert Hofmann's *Wonderchild* was born is often told as if it had been a mere accident. But to Albert himself, and to those who knew him well, it was no accident at all. Lysergic acid diethylamide, or LSD, was the offspring of both mystical intuition and scientific rigor, a twin force which seemed to propel Albert toward his discovery. Ergot and its alkaloids would become the fruitful yarn from which Albert would spin and weave his chemical tapestry, creating a beautiful artwork of compounds and medicines, which included the most potent spiritual catalyst humanity has ever known. This book represents a comprehensive record of Albert's groundbreaking research into the ergot alkaloids, finally translated into English.

Albert's foray into chemistry was not a matter of mere scientific curiosity; for him it was something of an existential compulsion. As a child, walking on a forest path on the Martinsberg near his home in Baden, Switzerland, Albert was enchanted by a profound mystical experience. He described it as if the veil of everyday vision had dissolved, illuminating the spring forest with a "ravishing radiance" and filling him with an "indescribable sensation of joy, oneness, and of blissful security." This and subsequent similar experiences compelled him to peer into the deeper fabric of nature and, seeking to uncover its mysteries, became the driving force of his life. The expectation of almost everyone who knew Albert was that he would choose a path in the humanities. Instead, he chose chemistry.

Most of all he was interested in working with plant and animal products, perhaps sensing that the most profound chemical structures would be discovered in close collaboration with nature. At Sandoz Laboratories, he was given the opportunity to study various medically significant plants whose active principles were unknown. Here, his curiosity would be captured by ergot, a lower fungus which had caused countless epidemics of convulsions and gangrene throughout history by infecting the ears of rye and other cereals which, when eaten in bread, poisoned the population, who thought it was the Witch's Curse. Despite its danger, it was known that the vasoconstrictive properties of ergot had played a prominent role in the pharmacopoeia of midwives in the Middle Ages, aiding childbirth and treating postpartum bleeding. Indeed, we have evidence that ergot was used for such purposes since ancient times by the Chinese, Greeks, and Mesopotamians. These societies appeared to have tamed the dangerous fungus by methods which remain a mystery to us today.

In our time, Albert would tame ergot himself through the rigorous methods of modern chemistry. When he first began his study of ergot, few of its alkaloids had been isolated, nor were their structures known. By the time of the first publication of this book in 1964, Albert and his colleagues had synthesized all the ergot alkaloids and identified their constituents. The building block of ergot alkaloids was identified as lysergic acid, which became the basis of a series of entirely novel compounds which Albert would develop in search of new or enhanced therapeutic properties. From both the natural alkaloids and their synthetic derivates would come a wellspring of invaluable medicines with effects primarily relating to blood circulation, such as preventing bleeding, reducing migraines, improving circulation and cerebral functioning, as well as stabilizing blood pressure.

And then there was his *Wonderchild*. By adding diethylamide to lysergic acid, Albert was expecting to produce a circulatory stimulant, similar to nicotinic acid diethylamide, with which he was familiar. This idea came to him during his lunchtime break, and he carried out the first synthesis of LSD on November 16, 1938. Disappointingly, when testing the compound, the pharmaceutical department found nothing of note, besides a modest uterotonic effect and a "restlessness" in the experimental animals, so it was discarded. But in April 1943, during another midday break, Albert writes: "The idea came to me in a strange way, again to synthesise lysergic acid diethylamide for further pharmacological testing. It was no more than a hunch which led me to take this unusual step—I liked the chemical structure of the substance."

During this re-synthesis on April 16, 1943, "chance had the opportunity to come into play. At the conclusion of the synthesis, I was overtaken by a very weird state of consciousness, which today one might call 'psychedelic.'" As he had worn protective visors and clothing and gloves to avoid any contact with the LSD he was synthesizing, he found this hallucination scientifically inexplicable; and it recalled to mind the visionary experiences of his childhood in the woods around Baden. The hallucinatory experience convinced him that he should perform a self-experiment with LSD, so on April 19, 1943 he consumed what he believed would be the smallest possible active dose of 250 micrograms, an amount which, to his amazement, would prove to be of alarming potency. What resulted was an experience of horror as he rode home on his bicycle while his consciousness transformed. When he got home, "the neighboring woman who brought me milk was no longer Mrs. R. but an evil, treacherous witch with a particolored grimace," and he feared the dissolution of his ego. But as his fear of insanity subsided, he was met with a beautiful inner mosaic of interwoven sensations that left him seeing the world in a fresh light.

In the afterglow of his experience, he realized he had come full circle: the spontaneous mystical experiences of his childhood had once guided him to study nature, and now his mastery of chemistry had yielded an elixir capable of taking him back to that mystical realm. Even more significantly, the creation of LSD meant that this spiritual awakening could be shared with anyone who wanted it.

In the two decades that followed, LSD would begin to revolutionize psychiatry and neuroscience, with its unparalleled ability to open up the psyche and allow the user to peer into the core of his or her own trauma and maladaptive patterns of thought and behavior. Through this illuminating process, LSD allowed users to change their setting to a more fertile ground where positive patterns could take root, altering the lens through which they saw the world. There followed an explosion of spirituality, artistic creation, and scientific innovation. In their LSD journeys, scientists and artists alike were confronted by a less constricted version of reality, the inner realm of the subconscious exploding into visionary insights. Personal computing, the double-helix structure of DNA, the greatest hits of the Beatles, and the iPhones in our pockets—all these discoveries, and countless more, owe their spark of inspiration to Albert's creation.

Albert would discover that LSD was not an anomaly, but rather the most potent in a family of related indole compounds. A fruitful collaboration with Gordon Wasson led to the discovery of the active principles of various shamanic "magic drugs" used by native Mexican tribes for thousands of years. Among them were the "magic mushrooms" that contained psilocybin and psilocin, compounds with structural similarities to LSD. Even more striking for Albert were the ololiuqui, morning glory seeds which were still being used in shamanic rituals, the name dating back to the Aztecs. To Albert's great surprise and delight, he discovered that the active principles of ololiuqui were ergot alkaloids, primarily lysergic acid amide and lysergic acid hydroxyethylamide, both closely related to LSD. Finding ergot alkaloids in higher plant species was considered so unlikely at the time that when Albert first presented his findings, they were largely dismissed by his colleagues.

After the publication of this book in 1964, Albert's work on ergot would further uncover one of the great secrets of our civilization. Alongside Gordon Wasson and Carl Ruck, Albert used his encyclopedic knowledge of the ergot alkaloids to show that the ancient Greek botanists of Eleusis had most likely mastered ergot's alchemical secrets, producing from it the potion known as the *kykeon*. This potion, and the associated rituals of the Eleusinian Mysteries, was the door through which the initiates experienced immortality and the heavenly world; it was the divine source of inspiration out of which the creative beauty of Western civilization would be born, including its art, democracy, science, and philosophy.

Decades after Wasson, Ruck, and Albert presented their thesis in *The Road to Eleusis*, archaeologists found ergot in a Greek ceremonial cup and in the dental cavities of an initiate to the Mysteries at Mas Castellar, Spain, in a temple devoted to Demeter and Persephone.

Sadly, the human animal's remarkable potential for brilliance has time and again been equaled by its potential for blindness. Eleusis and its mysteries would be destroyed by those who feared its power; the mid-wives of the Middle Ages would be persecuted as witches and burnt at the stake for their knowledge of ergot; and Albert's creation would be forced underground in the name of the War on Drugs. There it has remained for decades, buried and neglected along with all the knowledge we gathered during the productive years of science in the '50s and early '60s. But like the secrets of Eleusis, the secrets of science are not easily buried, and a slow revival is taking place.

When I finally met Albert for the first time in the '90s, I remember thinking that he was the happiest man I had ever met. I believe that I understand why. It must have been truly joyful to know that you have given the world such a wonderful gift. Although *Homo sapiens* may have become too clever for its own good, all may not be lost. There is a new Eleusis on the horizon, with a new science of how to better control our enhanced consciousness, inviting us to become a wiser, more compassionate *Homo sapiens* in closer partnership with Nature.

Amanda Feilding
May 2023

TRANSLATOR'S NOTE

My fascination with and curiosity about the experimentation and decades of accomplishments that resulted in the discovery of the structure and action of LSD brought me to the Novartis archive in Basel, home to Albert Hofmann's Sandoz laboratory notebooks, which contain glimpses into the wide scope of his chemical endeavors. Together with Hofmann's son Andreas, I visited their places of residence, including the family's last home "at the end of the world," Rittimatte, where Albert and his wife Anita lay buried, and which was, in Albert's words, "his most important discovery, along with LSD." Upon realizing that the laboratory notebooks had informed the publication of Albert Hofmann's classic, *Die Mutterkornalkaloide*—which precisely documents the extensive research that was required to allow a substance like LSD and other vital lysergic acid amide derivatives to be made—it was clear that this text deserved an English version.

Albert, an avid cyclist, world and life lover, natural products chemist, and mystic, working aptly in the same whereabouts as his favorite alchemist Paracelusus nearly half a millenium later, cared dearly that LSD, his "child," be regarded respectfully and with appreciation as "a sacred drug, whose contributions and potential for the world humanity has yet to fully understand." It is my genuine belief he would be delighted to see the still-widening medical, psychological, and spiritual interest in psychedelics that he spent so many years isolating, characterizing, researching, and developing.

Chemical names throughout the text mostly follow common names, and whenever possible keep the tone and style of the original in German even when different newer terms might sometimes also be in use now. I gratefully discussed plenty of nomenclature questions and received generous advice from colleagues whose help I want to acknowledge. Final decisions and responsibilities are my own.

For the contributions towards this English version of *Ergot Alkaloids* I give my thanks to and wish to acknowledge:

Andreas Hofmann, for generously and warmly welcoming me in Switzerland and sharing about their family (citations above of Albert's views are from our personal communications, written from my best memory and fragmented notes).

Thomas Szabo, an excellent chemist in the field who might have been a collaborator on the translation of this work but who died unexpectedly.

The developers of DeepL software, the capabilities of which immensely augmented, facilitated, and sped up the work.

Stephen Chapman for invaluable advice and shared passionate dive into nomenclature questions in thorough and comprehensive detail, always adding intriguing pieces of new information.

The whole Synergetic Press and Transform Press teams for making this project realizable in the first place, as well as highly enjoyable and smooth, in particular Doug Reil, Noelle Armstrong, and Allison Felus for their tremendous, constant help and kindness, and immense work.

Paul Freeman-Daley for composing the Bibliography and Author Index, shared inquiry in various nomenclature and chemistry-related questions, technical corrections and suggestions, and overall meticulous editing, helping to bring the book to its final version.

William Leonard Pickard for expert and thorough proofreading, investigating for and writing a beautiful Afterword, keeping note on progress of the project, all along advice, and beyond.

I dearly appreciate that this book is being copublished by Synergetic Press and Transform Press, founded by Ann and Alexander (Sasha) Shulgin, as Sasha and Albert were very good friends and rare, like-minded chemists.

Jitka Nykodemová
Philadelphia, Pennsylvania
Spring 2023

CHEMIST'S NOTE

At this writing we are celebrating the 80th anniversary of Albert Hofmann's spectacular discovery of the profound psychological effects of his LSD. Without question, no single molecule discovered by modern science has had the breadth of societal impact of this compound. For decades, LSD research and use was subject to international repression, born of lurid press coverage, fear, and political imperatives. Prohibition of LSD research robbed humanity of at least four decades of progress toward understanding of its potential benefits. Nevertheless, it would be challenging to quantify the impact of its efficacy in psychotherapy and its emergence as a sacrament and totem in youth culture. Today the ill winds of repression have shown evidence of change.

Hofmann certainly appreciated these impacts, which undoubtedly prompted his writing of *LSD—Mein Sorgenkind* (*LSD—My Problem Child*) in 1979, which candidly and sometimes uncomfortably addressed both the heights as well as the depths of experience produced by this material. However, the technical underpinnings of the discovery of LSD, detailed in this volume, have remained unavailable to the English-speaking world until now.

Some have described the discovery of LSD as a laboratory "accident," but nothing could be further from the truth. By the late 1930s, the alkaloids of the ergot fungus had been subject to intensive study as sources of valuable pharmaceutical remedies. Hofmann's earlier development of a rational method for releasing lysergic acid from more complex naturally occurring ergot alkaloids and the controlled addition of small functional groups had already produced ergonovine, an oxytocic that alone has saved millions of women's lives from death from postpartum hemorrhage. The variety of bodily targets and the physiological potency of ergot derivatives were not a mystery, yet the entirely unexpected psychological effects of LSD took Hofmann, and the scientific world, by complete surprise.   —

Reading this work, I was struck with the effort and ingenuity required to understand the detailed structure of the ergot alkaloids. Hofmann's group had, by today's standards, only the most rudimentary instrumentation to probe molecular secrets. They relied on painstaking and laborious separation of individual components of ergot, depending on methods only then recently developed, such as paper and thin-layer chromatography, to produce pure crystalline isolates of the many compounds present. These were then painstakingly broken into their component substructures that could then be isolated and crystallized. These were identified by comparison of physical properties with known or newly synthesized structures. Melting points, infrared and ultraviolet absorption spectra, or interaction with polarized light, were the only properties their instrumentality could address. There were no mass or nuclear magnetic resonance spectrometers, tools that today make structural identifications facile and rapid. The amount of labor involved to deduce even a single structure is staggering by today's standards, and as you will find, there are a host of structures presented here.

This monograph is a window into the past, into the chemical thought and diligence that formed the foundation of modern medicinal chemistry. For the reader interested in the many other aspects of ergot alkaloid history, production technology, and pharmacology, there are seminal volumes that elaborate further on these topics. The social history of ergot and the fungus' toxic effects on European health in the Middle Ages are detailed by Bové (1970). Berde and Schild (1978) extensively treated further developments in the pharmacology of ergot alkaloids. Békésy et al. (1973) describe the complex genetics of *Claviceps* species, both in natural infections and in the industrial fermentation environment now used to produce ergot alkaloids. Finally, Křen and Cvak (1999) thoroughly describe modern commercial production of ergot for pharmaceuticals.

To close, I am always reminded to place publications in time. When *Die Mutterkornalkaloide* was published, four years had elapsed since Ken Kesey's first experience with LSD as a volunteer in US Army projects later revealed to be part of the CIA's mind-control program, MKUltra. Alexander ("Sasha") Shulgin's discovery of highly potent and long-duration amphetamine psychedelics had not yet been published, let alone released into the nascent San Francisco drug scene. And it would be yet ten months before the Beatles made their debut on American TV. We have much to contemplate as we look forward to rehabilitating LSD and the pantheon of known psychedelics, bringing them back from near oblivion.

Paul F. Daley
Alexander Shulgin Research Institute
Berkeley, California
April 19, 2023

References

von Békésy, N., Spalla, C., Vining, L.C., Kobel, H., and Řeháček, Z. "Biosynthesis of Alkaloids" in Vaněk, Z., Hošťálek, Z., and Cudlínm (eds) *Genetics of Industrial Microorganisms. Vol. II Actinomycetes and Fungi.* Elsevier, Amsterdam-New York 1973.

Berde, B. and Schild, H.O. (eds.). *Ergot Alkaloids and Related Compounds. Handbook of Experimental Pharmacology.* Vol. 49. Springer-Verlag, Berlin-Heidelberg-New York 1978.

Bové, F. J. *The Story of Ergot. For Physicians, Pharmacists, Nurses, Biochemists, Biologists and Others Interested in the Life Sciences.* S. Karger, AG Basel-New York 1970.

Hofmann, A. *LSD—Mein Sorgenkind.* 1979 (English translation: *LSD—My Problem Child.*) McGraw-Hill, New York 1980.

Křen, V. and Cvak, L. (eds). *Ergot, The Genus Claviceps.* Harwood Academic Publishers, Amsterdam 1999.

PREFACE FOR REPUBLICATION

I am pleased and I am grateful to the publisher that the monograph *Ergot Alkaloids* is being published again in an unchanged reprint, as a historical document. The book was published in 1964 because by that time research in the field of ergot alkaloids had essentially come to an end.

This book comprehensively presents the chemistry and pharmacology of ergot alkaloids.

The chemistry of ergot alkaloids turned out to be not only a particularly interesting chapter in organic chemistry from a theoretical point of view but also a source of valuable medicines.

The following pharmaceutical preparations have emerged from ergot research: Gynergen, a migraine remedy; Methergine, a standard preparation in gynecology; Hydergine, a geriatric; and Dihydergot, a circulatory agent.

These medicines have remained valuable components of the medicinal treasury to this day.

However, the ergot compound lysergic acid diethylamide (LSD), which was clinically tested under the name Delysid, has not yet been able to develop into a medicine accessible to doctors. The pharmacological studies on LSD at the end of the book are the actual foundational publication of this substance, which is still being discussed worldwide as a miracle drug or as a satanic drug.

The new edition of *Ergot Alkaloids* is aimed at experts with an interest in history, but also at laypeople, as the book contains two easy-to-understand chapters on the botany of the ergot fungus and on the exciting history of ergot.

Albert Hofmann
June 2000

INTRODUCTION

The core of this monograph is a lecture given by the author under the title "Die Chemie der Mutterkornalkaloide" ("The Chemistry of Ergot Alkaloids") at the conference of the Deutschen Gesellschaft für Arzneipflanzenforschung (German Society for Medicinal Plant Research) in Tübingen in the fall of 1958. That conference was devoted to the topic of ergot. Lectures on the history and botany as well as on the pharmacology and therapeutic use of ergot alkaloids conveyed a comprehensive picture of this medicinal drug, which has kept the interest of physicians and natural scientists alive for centuries like hardly any other.

The specific active ingredients of ergot, the ergot alkaloids, are related in three respects to research areas of organic chemistry that have been heavily worked on: 1. They are fungal products. Research of fungal ingredients of fungi has taken on unprecedented proportions since the discovery of penicillin. 2. They are indole alkaloids. These take a prominent place in the framework of alkaloid research in recent decades. 3. The majority and the most important ergot alkaloids have peptide character. Peptide chemistry is becoming increasingly important in the field of hormones and antibiotics. In the peptide part of the ergot alkaloids, novel structural elements have been found which were previously unknown in organic chemistry.

The current interest in ergot alkaloids, due to their special chemical structure, is strengthened by the great importance of ergot preparations in the treasure trove of medicines, from which they have become indispensable. In addition to the classical use in obstetrics, ergot alkaloids or derivatives thereof have recently been increasingly used in internal medicine and in neurology and psychiatry.

Since the publication of the last summaries by *A. Stoll*[1-2], *A. L. Glenn*[3] and *H. Guggisberg*[4], important progress has been made in the field of ergot alkaloids: the total synthesis of the basic *core* of the ergot alkaloids, lysergic acid, was realized and its spatial structure elucidated; the peptide part of the peptide type ergot alkaloids has also been synthesized, its structure derived on the basis of degradation reactions, and its stereochemistry also determined; by linking the synthesized peptide part of ergotamine to the lysergic acid residue, the total synthesis of ergotamine was realized, whereby in principle making all ergot alkaloids totally synthetically accessible; another alkaloid pair, ergostine–ergostinine, belonging to a new series of peptide alkaloids, has been discovered; many new alkaloids have been found in the ergot of wild grasses, belonging to a new, simpler structural type; furthermore, a large number of other modification products of the natural ergot alkaloids have been produced, among which are those with remarkable pharmacological properties.

In this situation it seemed desirable to supplement the manuscript of the lecture mentioned at the beginning, in which a part of the more recent development had already been taken into account, with the latest research results mentioned above and to expand it into a monograph. The summarized presentation of the chemistry of ergot alkaloids in a monograph at the present time seems justified, because this field of research, which has occupied chemists for more than a hundred years, has today, as far as the chemistry of natural ergot alkaloids is concerned, reached a conclusion with the complete structural elucidation of the various alkaloid types and the realization of their total synthesis. On the other hand, studies on partial and total synthetic modification products of these alkaloids are continuing with the aim of further modifying their pharmacological qualities and finding new drugs. One line of work, which is also being pursued further, is concerned with the biogenesis of the ergot alkaloids.

The versatile pharmacological effects of ergot alkaloids and their derivatives and their great importance in therapy make it desirable to supplement the chemical section with appropriate chapters on pharmacology and the application of ergot alkaloids in medicine. To create a complete picture of the whole subject, an overview of the botany and cultivation of the ergot mushrooms, as well as the history of ergot, is given as an introduction.

I would like to express my best thanks to Dr. *F. Troxler* for reviewing the chemical chapters and to Dr. *M. Taeschler* for reviewing the pharmacological section of the manuscript. I am also indebted to Dr. *A. Brack* for advice in writing the section on the botany of the ergot fungus and for providing photographic material.

May this book, which takes into account the relevant literature up to the end of 1962, be useful to all fellow professionals interested in ergot alkaloids.

A. Hofmann

Basel

Spring 1964

ERGOT
ALKALOIDS

A. ON THE BOTANY OF THE ERGOT FUNGUS

I. <u>Claviceps purpurea</u> (Fries) Tulasne

The ergot of rye, officially called *Secale cornutum*, is from the biological point of view the permanent stage of the tubular fungus *Claviceps purpurea* (Fries) Tulasne. Figure 1 shows rye ears affected by ergot. The curved cones (sclerotia), colored in various shades from light brown to purplish brown, which protrude from the glumes instead of a normal rye grain, represent the overwintering form of the fungus. When the grain is mowed, or even before, the sclerotia fall to the ground where they remain through the winter until spring. With the onset of warm, humid weather the sclerotia begin to germinate. A prerequisite for germination is a cold period of several weeks during the winter dormancy. Large numbers of stalked, spherical fungal heads (stromata) consisting of hyphal bundles develop from the swollen ergot (Figure 2), the surface of which is covered with fine warts, each of which sits above a pit-shaped depression (perithecium). Figure 3 shows a section through a mushroom head (*Claviceps purpurea* of *Brachypodium pinnatum* [L.] Beauv.) with the perithecia clearly visible (original photo by *V. Grasso* "Le Claviceps delle Graminacee Italiane")[5]. Through a fine hole in the tip of each wart, the perithecia are open to the outside. At their base, clusters of elongated club-shaped tubes (asci) develop. Within each ascus are eight filamentous tubular spores (ascospores) formed by a sexual process (Figure 4). After ripening, the asci burst, the spores are ejected from the perithecia into the air and blown by the wind onto the stigmas of the rye flowers. In this way, the so-called primary infection of rye fields takes place. The ascospores germinate on the moist stigmas and form the mycelium of the secondary fruit form (sphacelia stage), which covers and then penetrates the

Figure 1. Ears of rye infected by the ergot fungus. (Photo by A. Brack)

Figure 3. Section through a mushroom head (*Claviceps purpurea* on *Brachypodium pinnatum* [L.] Beauv.). (Photo by V. Grasso)

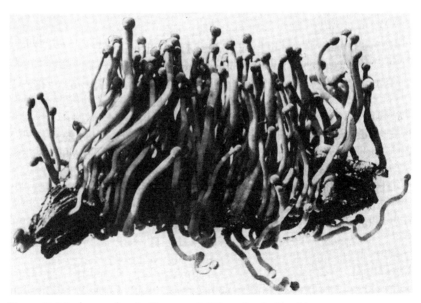

Figure 2. Mushroom heads (Stromata). (Photo by A. Brack)

2

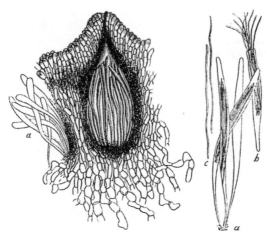

Figure 4. Schematic section through a perithecium. A) tubes (asci), B) open tube with ascospores, C) ascospores. (Photo from G. Barger, page 90)[6]

ovary. Sphacelia growth causes the host plant to secrete a sweet, reducing sugar-containing[7], sticky sap called honeydew (Figure 5). Embedded in this sap are innumerable oval spores, so-called conidia, formed from the sphacelia mycelium by asexual division. The honeydew now infects other flowering spikes by being transferred by insects or by touching the ears moving in the wind. This process is referred to as secondary infection. Over the course of a few weeks, the abundantly growing mycelium develops into the pseudoparenchyma, brown to purplish-brown in color in the outer layers, which consists of densely packed fungal cells with reserve substances, especially oils, thus forming the sclerotium, those dark cones known as ergot, which completes the development cycle.

II. Other species of Claviceps

Besides the practically most important *Claviceps purpurea* (Fr.) Tul., about 25

Figure 5. Ears of rye with honeydew. (Photo by A. Brack)

3

Table 1
Claviceps species according to Langdon
(supplemented until 1961)

Species	Synonym
C. annulata Langdon	
C. cinerea Griff.	
C. cynodontis Langdon	
C. diadema (Möller) Diehl	
C. digitariae Hansford	
C. flavella (Berk. and Curt.) Petch	*C. balansioides* Möller
	C. pallida Pat.
	C. patouillardiana P. Henn.
C. glabra Langdon	
C. grohii Groves	
C. hirtella Langdon	
C. inconspicua Langdon	
C. litoralis Kawatani	
C. lutea Möller	
C. maximensis Theis	
C. nigricans Tul.	
C. orthocladae (P. Herr.) Diehl	
C. paspali Stev. and Hall	*C. rolfsii* Stev. and Hall
C. phalaridis Walker	
C. platytricha Langdon	
C. purpurea (Fr.) Tul.	*C. microcephala* (Wallr.) Tul.
	C. sesleriae Stäger
	C. setulosa Quel.
C. pusilla Ces.	
C. queenslandica Langdon	
C. ranunculoides Möller	
C. sulcata Langdon	
C. tripsaci Stev. and Hall	
C. uleana P. Henn.	
C. viridis Padwick and Azmat.	
C. yanagawaensis Togashi	
C. zizaniae (Fyles) Pantidou	

other species of *Claviceps* are known, which mainly thrive on wild grasses. For their classification, differences in the formation of the stromata, perithecia, asci, ascospores and conidia are used. A recent overview of the different species of *Claviceps* and their differentiation was published by *R. F. N. Langdon*[8] (Table 1). Since the compilation of *R. F. N.*

4

Langdon two new species of *Claviceps* have been described, namely *C. phalaridis* Walker[9] and *C. zizaniae* (Fyles) Pantidou[10]. Lists of host plants of *Claviceps* species have been published by *T. Kawatani*[11], *V. Grasso*[12, 13], *J. C. Lindquist*[14], and *L. R. Brady*[15]. A given species of *Claviceps* may grow on different host plants. Conversely, a given plant is rarely susceptible to more than one *Claviceps* species. The alkaloid content of wild grass ergot is subject to very wide fluctuations[16].

While contamination of cereals with ergot in earlier times caused the dreaded manifestations of ergotism in humans, which will be discussed in more detail in the next chapter, ergot on forage grasses (e.g., on *Lolium perenne* in England, *Elymus canadensis* in North America, *Paspalum dilatatum* in Mississippi and Natal) often caused considerable damage to livestock[6]. Very recently, attention has also been drawn to damage caused by *Paspalum* ergot in Israel.

The ergot of wild grasses differs from the official ergot of rye mostly also in the alkaloid composition. While lysergic acid-based alkaloids predominate in rye ergot by far, ergot species containing mainly a simpler type of alkaloids, the alkaloids of the so-called clavine group, are also found on grasses.

The ergot of the tropical foxtail millet (*Pennisetum typhoideum* Rich.) has acquired a certain practical importance, because this fungus readily forms alkaloids even in saprophytic culture, but only those of the clavine type[17, 18]. In contrast, cultures of the species *Claviceps paspali* Stevens and Hall[19] could be obtained from the ergot of *Paspalum distichum* L., which produce lysergic acid derivatives in good yield in vitro[20].

III. Claviceps purpurea strains

Of the species *C. purpurea* (Fr.) Tul., which grows mainly on rye, there are several strains[21–23], which differ not only by certain morphological characteristics, but mainly by their ability to produce alkaloids. In addition to strains that produce more or fewer alkaloids, there are those that do not produce any alkaloids at all. Among the former, a distinction can be made between strains that produce almost exclusively a single alkaloid and those that can produce certain alkaloid groups. Since these properties are constant or hereditarily fixed in one and the same fungal strain, one may speak of different "chemical strains" within the species *Claviceps purpurea*. The host plant has a considerable influence on the amount and type of alkaloids in the sclerotia of *Claviceps purpurea*[24].

IV. The cultivation of the ergot mushroom

Most of the ergot of natural origin comes from Spain, Portugal and the Balkans, where good summers have seen harvests of over a hundred tons. The extensive use of ergot alkaloids as medicines has tremendously increased the demand for ergot drugs. On the other hand, as a result of careful seed selection and increased field care, natural infestation of rye fields with ergot has tended to decline. Today, ergot fungus is therefore artificially cultivated on a large scale, both on rye and in vitro in saprophytic culture.

1. Saprophytic culture

The cultivation of ergot fungus on artificial nutrient media has not yet led to any practical success as far as the production of medicinally used alkaloids is concerned. The fungus grows readily on suitable nutrient solutions and forms abundant mycelium both as a surface culture and after submergence, but only a few special strains are capable of producing alkaloids in the process. As M. *Abe* in Japan was the first to note, it is mainly some *Claviceps* species growing on wild grasses (*Agropyrum*, *Trisetum*, *Festuca*, *Elymus*, etc.) that are capable of forming alkaloids when cultured in vitro[25-27]. Quite considerable alkaloid yields could be obtained in the saprophytic culture of the ergot fungus of tropical foxtail millet (*Pennisetum typhoideum* Rich.)[17, 18]. However, these fungal strains produce mostly only clavine-type alkaloids, which have not yet acquired therapeutic significance. The conditions of alkaloid formation in submerged cultures have been studied by various authors[359-361, 366].

A. Stoll, A. Brack, A. Hofmann and *H. Kobel* succeeded for the first time in producing ergotamine in saprophytic cultures of a *Claviceps purpurea* strain of rye ergot[28]. Later, further successful experiments with strains of *Claviceps purpurea* became known, in which it was possible to demonstrate the formation of alkaloids contained in the official ergot (ergotamine–ergotoxine type)[29–35]. However, the alkaloid formation here is only small and inconsistent, so that an industrial utilization does not yet seem economical. A method for continuous alkaloid production with surface cultures of *Claviceps purpurea* was described by *D. Gröger* and *D. Erge*[362].

However, it was the observation of *F. Arcamone* et al., that another fungal species, *Claviceps paspali* Stevens and Hall, can be inoculated from ergot grown on the wild grass *Paspalum distichum* L., which is capable of forming considerable amounts of lysergic acid derivatives, especially lysergic acid amide and isolysergic acid amide, in submerged saprophytic culture[20, 36, 365]. Lysergic acid is obtained on an industrial scale by this process. The alkaloid yield could be improved by addition of racemic tryptophan, while it was slightly reduced by the addition of 5- or 6-methyltryptophan, and considerably lowered by 4-methyltryptophan[367].

2. Ergot culture on rye

For the production of the alkaloids of the ergotamine and ergotoxine groups, the artificial parasitic cultivation of the ergot on rye, which could be efficiently designed, is carried out industrially on a large scale. The first efficient method of artificial infection of rye fields was described by *N. v. Bekesy*[37]. Since then, this method has been developed by *A. Stoll* and *A. Brack*[38–40].

It is based on infecting the flowering rye ears with a conidia suspension obtained by in vitro culture, either by spraying or more effectively by injection. For large-scale industrial production, inoculation machines are now used that allow large fields to be efficiently infected by the injection method. The yield of such ergot crops is, of course, still dependent on the weather, but in humid, warm summers yields crops of several hundred kilos of ergot per hectare. Extensive ergot cultures of the company Sandoz A.G., Basel, exist in the Swiss Mittelland (Central Plateau). Also in other countries, e.g., Germany, Austria and in India, ergot is artificially cultivated. A great advantage of the artificial inoculation of rye fields is that by selecting suitable fungal strains, ergot can be produced that largely follows specific requirements in its alkaloid composition[24, 38, 41–43].

B. ON THE HISTORY OF ERGOT AND ITS ACTIVE SUBSTANCES

Ergot, like few other drugs, has a fascinating history in the course of which its role and importance has been reversed, first appearing as a feared poison carrier only to transform itself over time into a rich treasure trove of remedies, which even today is still not fully utilized.

The history of the ergot has found its classical representation in *G. Barger's* monograph "Ergot and Ergotism."[6] Those interested in the historical part should refer to this comprehensive study based on thorough research of the sources.

In antiquity, ergot does not seem to have played any role, because any references in ancient authors that could be clearly related to ergot are missing. This is understandable because, in the classical area around the Mediterranean, the host plant, rye, was hardly cultivated at that time.

Ergot first came to the attention of history in the early Middle Ages as the cause of epidemic mass poisonings. The disease, whose connection with ergot was not recognized at first, occurred in two characteristic forms, as gangrene (ergotism gangraenosus) and as convulsive ergotism (ergotism convulsivus).

The gangrene began with vomiting and diarrhea, tingling in the fingers and inflammatory manifestations accompanied by severe burning pains. Then, after a few days, the signs of gangrene began to appear. The limbs, first on the fingers and toes, began to turn blue-black and mummified. In cases of severe poisoning, the arms and legs could detach completely from the body without loss of blood. The gangrenous form of ergotism was referred to by disease names such as "mal des ardents," "ignis sacer," "holy fire."

In the convulsive form of ergotism, which began with symptoms similar to those of gangrenous ergotism, severe nervous disturbances were prominent. Painful muscular contractions occurred, namely of the extremities, which eventually turned into epilepsy-like convulsions. This manifestation was therefore also called "convulsive disorder" or "morbus epidemicus convulsivus."

An in-depth study of the occurrence of ergot epidemics from the early Middle Ages to more recent times has been conducted by R. *Kobert*[44].

From this we can see that a serious epidemic in 994 in Aquitaine and Limousin claimed about 40,000 victims, and that in the epidemic of 1129 in the Cambrai area 12,000 people died.

During the many ergot epidemics that struck the regions of Central Europe in the Middle Ages, the hospital fraternity of the Antonites in particular took care for those suffering from ergotism. This order, founded in 1093 in the south of France near Vienne, venerated St. Anthony as its patron saint. Since that time, the disease was also known as St. Anthony's fire. In a study of the medieval pictorial decoration of St. Anthony's church at Waltalingen (Switzerland), R. *Durrer*[45] relates the founding history of the order according to the Historia Antoniana of Aimar Falco in the Acta Sanctorum as follows:

> In the second half of the XL century, the relics of St. Anthony of Constantinople, who died in 356 at the age of 105, had been brought to St. Didier-La Mathe near Vienne (Dauphinee). Precisely then a neighboring nobleman, Gaston by name, had been stricken by the terrible disease of the holy fire and his young son Guerinus also succumbed to the contagion. Then they turned to the saint and vowed to consecrate their lives and possessions to him if they were cured. And St. Anthony justified the trust placed in him, having instructed the nobleman in a dream that he and his son should in the future care for those maimed by the disease at St. Didier. In response to the nobleman's fear that their fortune would not suffice for such an undertaking, he handed him a T-shaped staff which, planted in the ground, grew into a mighty tree, restoring the cripples with its shade and nourished them with its fruit. – Thus the founding legend of the Antönians [sic], whose insignia was the light blue T.

Already in 1095, at the Church Assembly of Clermont, which decided the First Crusade, Pope Urban II gave ecclesiastical approval to Gaston's foundation. In 1297, Pope Benedict VIII elevated the Antönians [sic] to an order of regulated canons with the Rule of St. Augustine.

An old woodcut from the Bayerische Staatssammlungen, Munich[46] shows St. Anthony surrounded by people suffering from ergotism (Figure 6).

Ergot as a cause of ergotism was first recognized by *Thuillier,* personal physician to the Duke of Sully, on the occasion of an epidemic in Sologne (1630) and confirmed by feeding experiments on poultry. Sologne, south of the Loire near Orleans, was for centuries a notorious focus of gangrenous ergotism.

In most European countries and also in certain areas of Russia, the epidemic-like occurrence of ergot poisoning has been recorded until modern times. Ergotism mostly occurred in connection with war and famine. It is probable that the outbreak of ergot epidemics after wet summers, favorable for the growth of the fungus, was promoted by malnutrition and a certain avitaminosis of the poor class of the population.

Figure 6. St. Anthony, surrounded by people suffering from ergotism. (Original in the possession of the National Graphic Arts Collection, Munich.)

In this context, reference should be made to a study by *E. Mellenby*[47] showing that in the case of vitamin A deficiency, the susceptibility to ergotism is greater. With the general improvement of the nutritional situation, with the improvement of agriculture, and after the realization, obtained in the 17th century, that bread containing ergot was the cause of ergotism, the frequency and magnitude of ergot epidemics decreased steadily. The last major epidemic affected certain areas of Russia in 1926/27. Between Kazan and the Urals, 11,000 people fell ill with ergot poisoning, 93 of whom died.

The first description of ergot and at the same time the first mention of its medicinal use is found in the herbal book of the Frankfurt city physician *Adam Lonitzer* (Lonicerus) from 1582. On page 285 is the text, which is reproduced in facsimile in Figure 7.

The first illustration of ergot is considered to be a woodcut in *Caspar Bauhin*'s Theatrum Botanicum, printed in Basel in 1658. While the drug is called by *Lonitzer* as "Kornzapfen", referred to as "Secale Luxurians" by *Bauhin*, the name "ergot" is first encountered by *Mauritius Hoffmann* in his "Flora Altdorffina," 1662.

> Nota : Von den Kornzapffen / Latinè, Claui Siliginis: Man findet offtmals an den ähren deß Rockens oder Korns lange schwartze harte schmale Zapffen / so beneben vnnd zwischen dem Korn / so in den ähren ist / herauß wachsen / vñ sich lang herauß thun / wie lange Neglin anzusehen / seind innwendig weiß / wie das Korn / vnd seind dem Korn gar vnschädlich.

> Solche Kornzapffen werden von den Weibern für ein sonderliche Hülffe vnd bewerte Artzney für das auffsteigen vnd wehethumb der Mutter gehalten / so man derselbigen drey etlich mal einnimpt vnd isset.

Figure 7. First description of ergot in Adam Lonitzer's Kräuterbuch, Frankfurt 1582. (From G. Barger "Ergot and Ergotism")[6]

Although ergot has been used by midwives as an oxytocic since ancient times, as can be seen from the above quotation from Lonicerus, and in the 18th century its use as *pulvis ad partum* was also documented by some physicians, it was not until 1808 that this drug found its way into official medicine on the basis of a work by the American physician *John Stearns* entitled "Account of the Pulvis Parturiens, a Remedy for Quickening Childbirth."[48] The usefulness of the application of ergot as contraceptive did not remain however undisputed[49]. Partly failures were reported, partly increased infant mortality was pointed out, and *pulvis ad partum* was called virtually *pulvis ad mortem* as far as the child was concerned. Such contradictory results are readily understandable today, now that one knows that ergot of different provenance shows large variations in the active substance content, indeed that there are ergot varieties which contain no alkaloids at all. The great danger for the child, which lay above all in the unreliable, too high dosage, was recognized at an early stage, and ergot use was limited to stopping postpartum hemorrhage. This has remained the main indication for ergot alkaloids in obstetrics to this day.

After ergot was included in various pharmacopoeias in the first half of the 19th century, the first chemical work on isolating the active components of this drug also began.

The first pharmaceutical-chemical study worth mentioning was published in 1816 by *M. Vauquelin*[50]. However, he did not succeed, nor did the numerous researchers who dealt with this problem in the next 100 years, in identifying the actual vehicles of the therapeutic effect of ergot and in defining it chemically, and so the views on the chemical nature of the ergot active substances changed according to the respective, often contradictory, findings. In 1875, *Ch. Tanret* succeeded for the first time in isolating a crystallized alkaloid, ergotinine[51], which, however, proved to be ineffective. Even when *G. Barger* and *F. H. Carr*[52] in 1907 obtained from ergot an effective and, as was later shown, a complex alkaloid preparation, ergotoxine, and *F. Kraft* hydroergotinine[53, 54], identical with ergotoxine, it had not yet been decided whether the carriers of ergot action were alkaloidal in nature. During the pharmacological testing of ergotoxine, *H. H. Dale*[55] noted its contractile effect on the uterus and also made the significant discovery, fundamentally important for the present therapeutic application of ergot alkaloids, that this substance has a specific antagonistic effect to adrenaline on the autonomic nervous system. However, these preparations disappointed when used in the clinic, and there was a tendency to believe that not alkaloids, but simpler plant bases such as the biogenic amines, tyramine and histamine, as well as acetylcholine, which were detected in ergot extracts, were the carriers of the uterine contractile effect. Because of the high toxicity of these first alkaloid preparations, which was already expressed in the name "ergotoxine," it was even recommended to remove the alkaloids from ergot extracts for therapeutic use[53].

A turnaround in the therapeutic evaluation of ergot ingredients occurred when *A. Stoll* succeeded in 1918 in isolating for the first time a uniform, crystallized alkaloid, which he called ergotamine[56–61] and which exhibited the full effects of the ergot drug in pharmacological tests and in the clinic[62], and soon found extensive use in therapy. With this fundamental discovery, a new, contemporary phase of ergot research began. It was now certain that alkaloids are the actual carriers of ergot action, and further research from then on concentrated on the alkaloid constituents of the drug.

In parallel with further chemical research into ergot alkaloids, which is summarized in the chronologically arranged table in the next section and the results of which are presented in detail in the following chapters, the number of pharmacological studies with these alkaloids and their derivatives has increased and, accordingly, also the range of applications of ergot ergot preparations (cf. Section D).

Ergot growing in the rye fields as a result of natural infection was soon insufficient for a now sought-after drug for production of pharmaceutical preparations, and methods for the artificial inoculation of rye

14

were developed. Furthermore, in the search for high-yielding sources of ergot alkaloids, chemical races of the *Claviceps* fungus have been bred out and successful studies have been conducted to find strains of the fungus that produce alkaloids in vitro on an industrial scale (see Section A, IV).

This is, in brief, the history of ergot and its active ingredients, which were once the cause of mass poisoning and have now become sought-after starting materials for the production of valuable remedies.

C. CHEMISTRY OF THE ERGOT ALKALOIDS

The chemistry of ergot alkaloids has been elucidated in all aspects to a great extent and now provides a nicely completed picture. In order not to have to disrupt this coherence in the following presentation with historical information, it is preceded by an overview of the chronological sequence of ergot alkaloid research.

I. Historical overview of the various phases of chemical research of ergot alkaloids

1. Isolation and purification of the alkaloids

The great sensitivity of ergot alkaloids to chemical agents, atmospheric oxygen and light and their easy isomerization made their isolation and purification a difficult task, which has engaged chemists for more than a hundred years, and which could only be fully resolved in the last decades.

1816: First noteworthy pharmaceutical-chemical investigation of ergot by *M. Vauquelin*[50].

1875: Isolation of the first crystallized alkaloid preparation, ergotinine, by *C. Tanret*[51].

1906: The first pharmacologically active alkaloid preparation, ergotoxine, is described by *G. Barger* and *F. H. Carr*[52, 63] and simultaneously by *F. Kraft*[53, 54], who calls it hydroergotinine. The alkaloid base was amorphous, but gave a crystallized phosphate.

1918: Isolation of the first chemically consistent, fully active ergot alkaloid, ergotamine, by *A. Stoll*[56–61].

1935: Isolation of a water-soluble, specifically uterus-active alkaloid by four different laboratories. *H. W. Dudley* and *C. Moir*[64] named the new alkaloid ergometrine, *A. Stoll* and *E. Burckhardt*[65, 66] ergobasine, *M. S. Kharasch* and *R. R. Legault*[67] ergotocine and *R. M. Thompson* ergostetrine[68].

1937: Another alkaloid, ergosine, is purified by *S. Smith* and *G. M. Timmis*[69].

1937: Ergocristine, another pure alkaloid, is isolated by *A. Stoll* and *E. Burckhardt*[70] from Iberian ergot.

1943: Ergotoxine, which had previously been considered a single homogenous alkaloid, is resolved by *A. Stoll* and *A. Hofmann*[71] into three components, the already known ergocristine and two new alkaloids, ergocryptine and ergocornine.

1951: From Japanese grass ergot, the first representative of a new type of alkaloid, agroclavine, is isolated by *M. Abe*[25, 26].

1952: Elymoclavine, a second clavine-type alkaloid, is discovered by *M. Abe* et al.[27].

1954: Isolation of penniclavine from the ergot of African foxtail millet *(Pennisetum typhoideum* Rich.) by *A. Stoll* et al.[17].

1954: Festuclavine is isolated from ergot of *Festuca rubra* L. by *M. Abe* and *S. Yamatodani*[72].

1955: Molliclavine is isolated from the ergot of *Elymus mollis* Trin. by *M. Abe* and *S. Yamatodani*[73].

1956: Pyroclavine and costaclavine, two new alkaloids from ergot of *Agropyrum semicostatum* Nees, are described by *M. Abe* et al.[74].

1957: *A. Hofmann* et al.[18] isolate isopenniclavine, setoclavine, isosetoclavine and chanoclavine, four other alkaloids of the clavine group, from the fungal strain of *Pennisetum typhoideum* Rich.

1959: *M. Abe* et al.[75] describe ergosecaline and ergosecalinine, a peptide-type alkaloid pair found in Iberian ergot.

1960: Lysergene is isolated by *S. Yamatodani*[76] from saprophytic cultures of an *Elymus*-type fungus.

1961: *M. Abe* et al.[77] describe the isolation of lysergine from *Agropyrum*-type saprophytic cultures and of lysergol from an *Elymus* fungal strain. Lysergol is simultaneously detected by *A. Hofmann*[78] in the seeds of *Rivea corymbosa* (L.) Hall. f.

1961: Fumigaclavine A and fumigaclavine B are isolated from *Aspergillus fumigatus* Fres. by *J. F. Spilsbury* and *S. Wilkinson*[79].

1962: Ergostine and ergostinine, a new peptide-type alkaloid pair is found by *W. Schlientz* et al.[80] in small quantities in rye ergot.

2. Structural elucidation and synthesis of ergot alkaloids

1932: Isolation of the first crystallized, characteristic cleavage product of ergot alkaloids, later proven to be isolysergic acid amide, by *S. Smith* and *G. M. Timmis*[81].

1934: Lysergic acid, a product of alkaline hydrolysis, is identified by *W. A. Jacobs* and *L. C. Craig* as the basic building block of ergot alkaloids[82, 83].

1935: Ergometrine (ergobasine) is formulated on the basis of its hydrolysis products by *W. A. Jacobs* and *L. C. Craig* as lysergic acid (+)-propanolamide-(2)[84].

1935: Evidence of a peptide-like residue in ergotoxine and ergotamine by *W. A. Jacobs* and *L. C. Craig*[85–87].

1936: Description of isolysergic acid by *S. Smith* and *G. M. Timmis*. The lysergic acid–isolysergic acid isomerism is recognized as the cause of the paired occurrence of the ergot alkaloids[88].

1937: First partial synthesis of an ergot alkaloid, ergobasine (ergometrine) from lysergic acid and L-(+)-2-aminopropanol by *A. Stoll* and *A. Hofmann*[89].

1938: Establishment of a structural formula for lysergic acid by *W. A. Jacobs* and *L. C. Craig*, which later proved to be correct except for the position of a double bond[90].

1945: Total synthesis of racemic dihydrolysergic acid by *F. C. Uhle* and *W. A. Jacobs*[91].

1949: Elucidation of the lysergic acid isomerism and final formulation of lysergic acid by *A. Stoll, A. Hofmann* and *F. Troxler*[92].

1950: Total synthesis of the optically active dihydrolysergic acids by A. Stoll, J. Rutschmann, and W. Schlientz[93].

1951: Establishment of the complete constitutional formula of the peptide alkaloids of the ergotamine and ergotoxine group (cyclol formula) by A. Stoll, A. Hofmann, and Th. Petrzilka[94].

1954: Total synthesis of lysergic acid by E. C. Kornfeld et al.[95, 96].

1954: Elucidation of the stereochemistry of lysergic and dihydrolysergic acids by A. Stoll et al.[97].

1959: Determination of the absolute configuration of lysergic acid by physical means by H. G. Leemann and S. Fabbri[98], which was later confirmed by chemical coupling by P. Stadler and A. Hofmann[99].

1961: Total synthesis of ergotamine by A. Hofmann, A. J. Frey, and H. Ott[100].

II. Structural types of ergot alkaloids

The ergot alkaloids belong to a large important class of indole alkaloids. The indole moiety is built into a tetracyclic ring system called ergoline, which represents the characteristic backbone of all ergot alkaloids[101]. Ergoline has a special place among the various types of indole alkaloid structures in that a ring is attached to the 4-position of the indole system. In all other indole alkaloids known to date, further rings are attached to the 1,2,3- or 7-position.

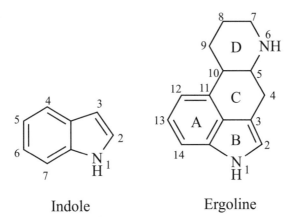

Indole   Ergoline

The natural ergot alkaloids can be divided into two main groups based on their chemical structure.

Main group A: Acid amide-like derivatives of lysergic and isolysergic acid with the general formula A.

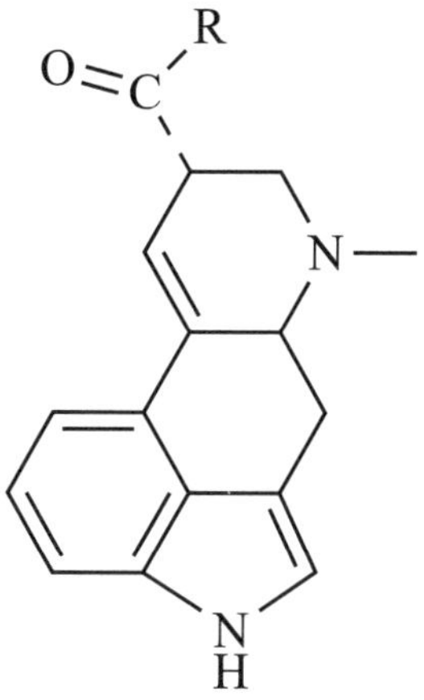

A

R = cyclic peptide residue: **Peptide type**

Ergotamine – ergotaminine

Ergosine – ergosinine
} Ergotamine group

Ergocristine – ergocristinine

Ergocryptine – ergocryptinine
} Ergotoxine group

Ergocornine – ergocornine

Ergostine – ergostinine

Ergosecaline – ergosecalinine

R = L-2-aminopropanol: **Alkanolamide type**

Ergobasine – ergobasinine

R = NH_2: **Amide type**

Ergine – erginine

R = $NHCHOHCH_3$: **Carbinolamide type**

d-Lysergic acid methyl carbinolamide

21

Main group B: Group of clavines with the general formula B.

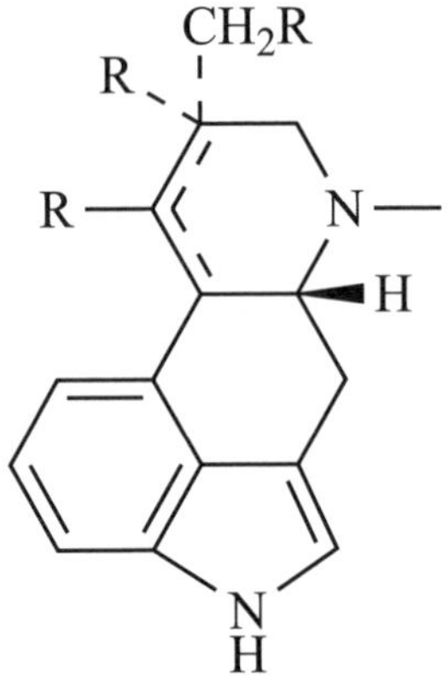

B

R = H or HO

A special scaffold within the clavine group is represented by the alkaloid chanoclavine, in which the ring D between the nitrogen atom and the C7 atom is open.

III. The group of lysergic acid alkaloids

In the following section, the individual alkaloids of the lysergic acid group are described in more detail.

The alkaloid pairs ending in "-ine" are based on lysergic acid, the alkaloids whose names end in "-inine" are based on isolysergic acid. The isomeric alkaloids are interconvertible. The lysergic acid alkaloids rotate left or weakly right and are distinguished by high pharmacological activity. The isomeric alkaloid pairs derived from the isolysergic acid are strongly dextrorotatory and are only little active in most pharmacological tests.

By gentle processing of the completely fresh material, one usually obtains predominant amounts of the levorotatory alkaloids[102]. From an old material, or during improper isolation, mixtures are obtained in which the dextrorotatory, ineffective forms predominate. Especially in alkaline environments, the genuine lysergic acid alkaloids are rapidly isomerized.

1. The individual alkaloids of the lysergic acid group

With the isolation of ergotamine by *A. Stoll* from Swiss ergot in 1918, the first chemically truly uniform, fully effective ergot alkaloid was found[57, 61]. Ergotamine was also the first ergot alkaloid to find widespread, versatile medical use (cf. Section D).

Ergotamine

$C_{33}H_{35}O_5N_5$ (m.w. 581.7)

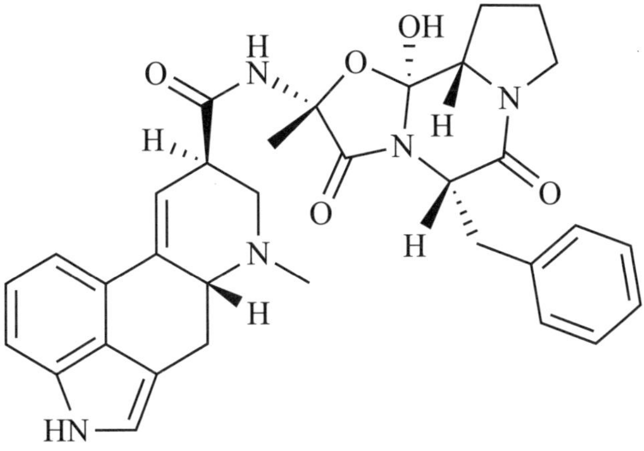

The large demand for ergotamine could soon no longer be met from naturally grown ergot, so that it was necessary to turn to artificial, mechanical infection of rye crops on an industrial scale (cf. Section A, IV/2). By far the largest part of this ergot alkaloid, which is still the most important of all ergot alkaloids in terms of medicinal use, is obtained by Sandoz AG, Basel, from ergot cultivated by this process in the Swiss Mittelland.

For the isolation and purification of ergotamine from the ergot drug, which can be carried out by the methods generally used for the extraction of alkaloids, a procedure developed by *A. Stoll*, which takes into account the easy decomposability of this alkaloid and which is also suitable for industrial extraction, has proved particularly successful[57, 61]. It is based on fixing the ergotamine base in the coarsely ground drug with aluminum sulfate, after which the abundant ergot oil and other accompanying substances can be removed with the same solvent with which the alkaloid is subsequently extracted, namely with benzene. After alkalization with ammonia, the ergotamine can be extracted with benzene from the drug pretreated in this way in such a pure form that it crystallizes out when the extract is concentrated.

The yields from naturally grown ergot of different origin are 1–2 g ergotamine (base) per kg ergot and can be increased by artificial infection with selected fungal strains.

Ergotamine crystallizes particularly easily and typically from 90% aqueous acetone in straight-cut, polyhedral, crystallization solvent-containing prisms of the composition $C_{33}H_{35}O_5N_5 \cdot 2\ CH_3COCH_3 \cdot 2\ H_2O$, which melt indistinctly at 180°. The crystals weather when kept in air for a long time and completely release the crystallization solvent in vacuum at elevated temperature. From 800 parts of boiling benzene, ergotamine crystallizes on cooling in long, thin prisms that melt at 212–214° with decomposition. At room temperature, the solvent-free alkaloid dissolves in approximately 300 parts of ethanol, 70 parts of methanol, and 150 parts of acetone. It is slightly soluble in chloroform and practically insoluble in water.

$pK^* = 5.6$ (*in 80% aq. methyl cellosolve*)
$[\alpha]_D^{20} = -160°$; $[\alpha]_{5461}^{20} = -192°$ ($c = 1.0$ *in CHCl₃*)
$[\alpha]_D^{20} = -12.7°$; $[\alpha]_{5461}^{20} = -8.6°$ ($c = 1.0$ *in pyridine*)

IR spectrum see Section V, 3b, Figure 15a.

Keller color reaction: blue (see Section V, 2a).

Salts of ergotamine:

Tartrate $(C_{33}H_{35}O_5N_5)_2 \cdot C_4H_6O_6 \cdot 2\ CH_3OH$. Thick rhombic plates, m.p. 203° (dec.). This salt, which is relatively soluble in water, is the most common form of ergotamine used in pharmaceutical preparations (Gynergen®, Neo-Gynergen®, Bellergal®, etc.) and is also included in pharmacopoeias (e.g., USP XVI and ÖAB 9).

Hydrogen maleate. From methanol in rhombic crystalline solvent-containing platelets, m.p. 195–197° (dec.).

Hydrochloride. From 90% alcohol obliquely truncated, long, thin plates and prisms, m.p. 212° (dec.).

Phosphate. From 90% alcohol in thin, hexagonal, crystal solvent-containing platelets united into druses, m.p. 200° (dec.).

Sulfate. This very sparingly soluble salt crystallizes from 70 parts of 80% aqueous methanol in rhomboid plates, m.p. 207° (dec.).

Methanesulfonate. This relatively easily soluble salt crystallizes from alcohol in thin, hexagonal platelets, m.p. 210° (dec.).

Ergotaminine
$C_{33}H_{35}O_5N_5$ (m.w. 581.7)

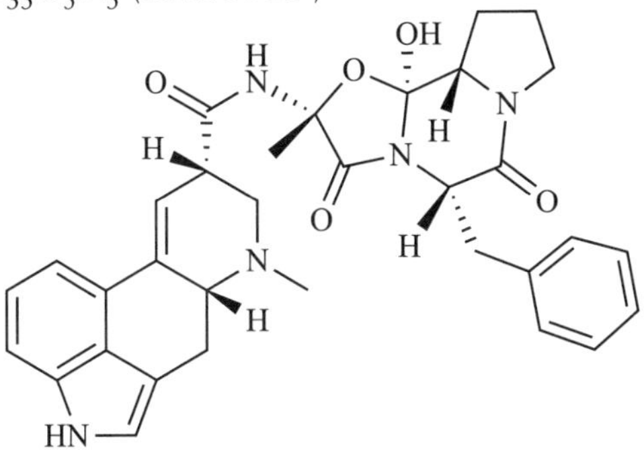

Ergotaminine is present in fresh ergotamine-containing ergot drugs only in small quantities, but is formed in considerable amounts from the pharmacologically active isomer during aging, especially if the drug is not stored completely dry.

Due to its poor solubility, ergotaminine crystallizes out of the equilibrium, which is rapidly established between ergotamine and ergotaminine in hydroxyl-containing solvents, thus disturbing the equilibrium until the virtually complete conversion of ergotamine into ergotaminine. Conversely, the extremely low solubility of ergotamine sulfate can be used to continuously crystallize the resulting ergotamine from a solution of ergotamine in sulfuric acid–containing glacial acetic acid, whereby a practically complete rearrangement of ergotaminine into ergotamine can be achieved[61].

Ergotaminine, like most derivatives of isolysergic acid, is pharmacologically practically ineffective. Only a certain adrenolytic activity on the guinea pig seminal vesicle was found[103].

Ergotaminine is extremely poorly soluble in most solvents. It dissolves only in about 1,500 parts of boiling methanol or in 1,000 parts of boiling ethanol. Ergotaminine is relatively soluble in pyridine or glacial acetic acid. From a heat-saturated solution in methanol it precipitates on cooling in typical rhombic, thin platelets with truncated corners, containing no crystalline solvent, m.p. 241–243° (dec.).

$pK^* = 5.6$ (*in 80% aq. methyl cellosolve*)
$[\alpha]_D^{20} = +369°; [\alpha]_{5461}^{20} = +462°$ ($c = 0.5$ *in chloroform*)
$[\alpha]_D^{20} = +397°; [\alpha]_{5461}^{20} = +497°$ ($c = 0.5$ *in pyridine*)

IR spectrum see Section V, 3b, Figure 15b.

Keller color reaction: blue (see Section V, 2a).

Ergotamine forms no stable salts.

Aci-ergotamine
$C_{33}H_{35}O_5N_5$ (m.w. 581.7)

Aci-ergotamine forms at room temperature upon prolonged storage of ergotamine in acidic solution. It differs from the latter by the steric arrangement in the peptide moiety. When acidic solutions are boiled, an equilibrium of ergotamine, aci-ergotamine, ergotaminine and aci-ergotaminine is quickly formed. Due to its amphoteric character, aci-ergotamine can be easily separated from ergotamine chromatographically or by partitioning the organic and alkaline aqueous phases[104].

Due to the rearrangement in the peptide part of ergotamine yielding aci-ergotamine, the pharmacological efficacy is largely lost. Only a weak uterotonic effect on the rabbit uterus in situ can still be observed with aci-ergotamine[103].

Aci-ergotamine crystallizes from methanol in needles of m.p. 185–187° (dec.).

$pK^* = 5.5$ (*in 80% aq. methyl cellosolve*)
$[\alpha]_D^{20} = -32°$ (*c = 1.2 in pyridine*)

Aci-ergotaminine
$C_{33}H_{35}O_5N_5$ (581,7)

In acidic solution, aci-ergotaminine is in equilibrium with ergotaminine, aci-ergotamine and ergotamine[104]. Aci-ergotaminine may be considered biologically practically ineffective[103].

From methanol/ether in fine needles of m.p. 203° (dec.)

Ergosine
$C_{30}H_{37}O_5N_5$ (m.w. 547.6)

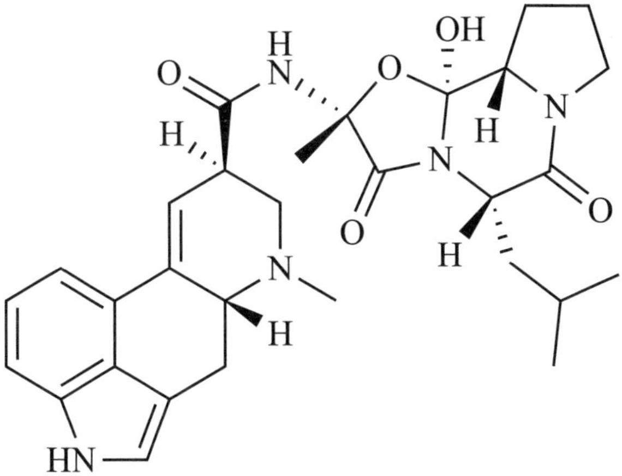

Ergosine, which was first isolated and described by *S. Smith* and *G. M. Timmis* in 1937[69], has not yet found any medical application.

It occurs in small quantities in Iberian ergot, which contains the alkaloids of the ergotoxine complex as its major constituent.

From ethyl acetate, rectangular plates, m.p. 220–230° (dec.)

$pK^* = 5.5$ (*in 80% aq. methyl cellosolve*)
$[\alpha]_D^{20} = -183°$; $[\alpha]_{5461}^{20} = -220°$ (*c = 1.0 in chloroform*)
$[\alpha]_D^{20} = -8°$; $[\alpha]_{5461}^{20} = -1°$ (*c = 1.0 in pyridine*)

IR spectrum see Section V, 3b, Figure 15c.

Keller color reaction: blue (see Section V, 2a).

Salts of ergosine:

Hydrochloride: crystallizes from acetone in crystal solvent-containing plates, m.p. 235° (dec.).

Methanesulfonate: From methanol in needles merged into clusters, m.p. 217–218° (dec.)

Ergosinine

$C_{30}H_{37}O_5N_5$ (m.w. 547.6)

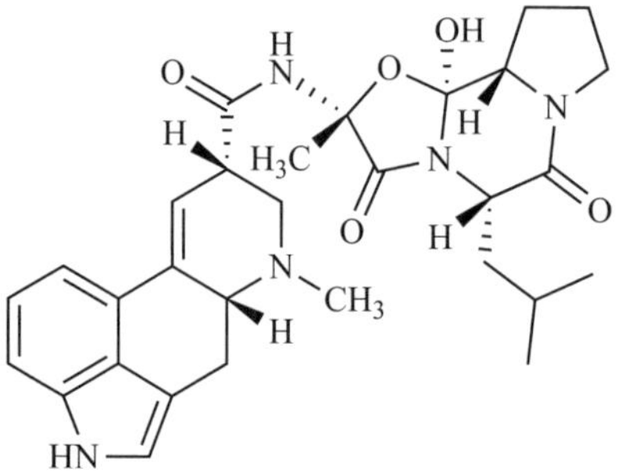

Ergosinine can be reversibly rearranged into ergosine by alkaline or acidic treatment[69].

From 90% acetone in obtuse prisms, m.p. 228° (dec.).

$pK^* = 5.5$ (*in 80% aq. methyl cellosolve*)

$[\alpha]_D^{20} = +420°$; $[\alpha]_{5461}^{20} = +522°$ ($c = 1.2$ *in chloroform*)

IR spectrum see Section V, 3b, Figure 15d.

Keller color reaction: blue (see Section V, 2a).

Ergocristine

$C_{33}H_{39}O_5N_5$ (m.w. 609.7)

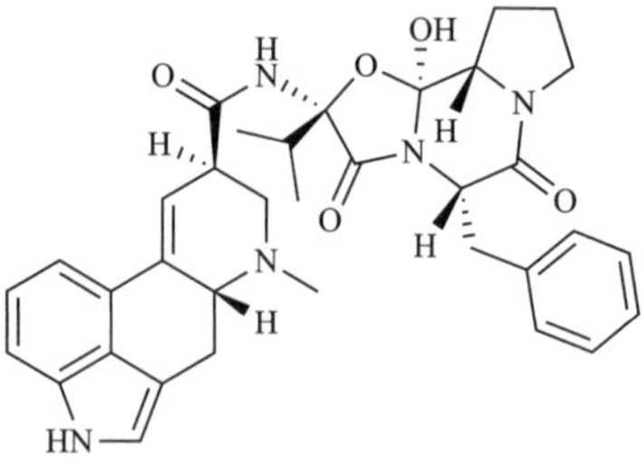

Ergocristine was isolated in 1937 by *A. Stoll* and *E. Burckhardt*[70] from Spanish and Portuguese ergot and was later recognized, along with ergocornine and ergocryptine, as a constituent of ergotoxine first described 30 years earlier[52]. Ergotoxine preparations of different origin were separated into the single alkaloid components via the salt with di-(*p*-toluyl)-L-tartaric acid and were proven to be variable mixtures of ergocristine, ergocryptine and ergocornine[71].

Ergocristine, as the main alkaloid, is also contained in ergot of German and North American origin, along with ergotamine. In the form of its strongly sympatholytic dihydro derivative, it is used medicinally, e.g., as a component of "Hydergine"® (cf. Section D).

Ergocristine is characterized by its crystallization from acetone, from which it separates in planar, obliquely truncated prisms, that stubbornly retain 1 mole of crystallization acetone and that melt indistinctly with decomposition at 160–175°. Ergocristine dissolves in 40 parts boiling benzene and crystallizes from it in elongated, rectangularly confined plates, which, on prolonged standing, rearrange themselves into roof-shaped, massive prisms. The alkaloid is very soluble in methanol, ethanol and chloroform.

$pK^* = 5.5$ (*in 80% aq. methyl cellosolve*)
$[\alpha]_D^{20} = -180°$; $[\alpha]_{5461}^{20} = -217°$ (*c = 1.0 in chloroform*)
$[\alpha]_D^{20} = -108°$; $[\alpha]_{5461}^{20} = -125°$ (*c = 1.0 in pyridine*)

IR spectrum see Section V, 3b, Figure 15c.

Keller color reaction: blue, after a few seconds turns olive green (see Section V, 2a).

Salts of ergocristine:

Hydrochloride, from alcohol/ether elongated sheets, m.p. 205° (dec.).

Phosphate, from alcohol hexagonal plates, m.p. 195° (dec.).

Ethanesulfonate, from acetone in elongated hexagonal sheets, m.p. 207° (dec.).

Ergocristinine

$C_{33}H_{39}O_5N_5$ (m.w. 609.7)

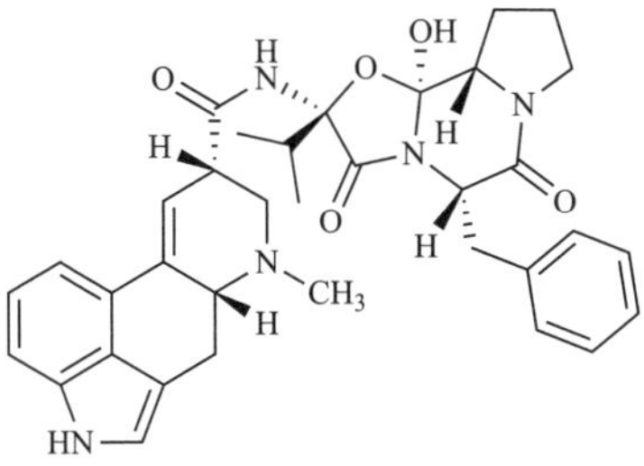

The reconversion of ergocristinine into ergocristine is achieved by boiling in alcoholic phosphoric acid[70].

Ergocristinine[70, 71] crystallizes from alcohol, in which it is rather difficult to dissolve, in long, thin prisms of m.p. 228° (dec.). The alkaloid is readily soluble in chloroform and in ethyl acetate.

$pK^* = 5.0$ (*in 80% aq. methyl cellosolve*)
$[\alpha]_D^{20} = +366°$; $[\alpha]_{5461}^{20} = +460°$ (*c = 0.7 in chloroform*)
$[\alpha]_D^{20} = +462°$; $[\alpha]_{5461}^{20} = +582°$ (*c = 1.0 in pyridine*)

IR spectrum see Section V, 3b, Figure 15f.

Keller color reaction: blue, after a few seconds turns olive green (see Section V, 2a).

Ergocryptine

$C_{32}H_{41}O_5N_5$ (m.w. 575.7)

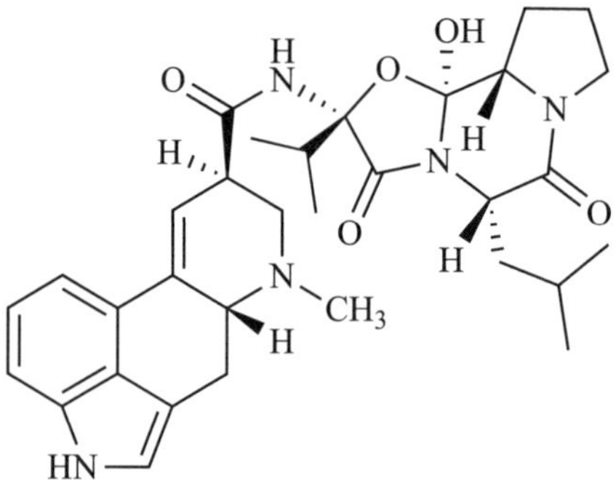

Ergocryptine was isolated as a component of the ergotoxine complex via the salt with di-(*p*-toluyl)-L-tartaric acid. As the main alkaloid, ergocryptine has been found in the ergot of the grass *Elymus mollis* Tri[27] growing in Japan and in the ergot of the South American grass *Carex arenaria* (Cyperaceae)[105]. In the form of the dihydro derivative, it is used medicinally as a component of "Hydergine"®.

Ergocryptine[71] crystallizes from the concentrated methanolic solution in straight-cut prisms of m.p. 210–212° (corr.). The alkaloid is also readily soluble in ethanol, acetone and chloroform. From benzene it crystallizes in the same crystal modifications as ergocristine and ergocornine.

$pK^* = 5.5$ (*in 80% aq. methyl cellosolve*)
$[\alpha]_D^{20} = -191°$; $[\alpha]_{5461}^{20} = -228°$ (*c = 1.0 in chloroform*)
$[\alpha]_D^{20} = -117°$; $[\alpha]_{5461}^{20} = -138°$ (*c = 1.0 in pyridine*)

IR spectrum see Section V, 3b, Figure 15g.

Keller color reaction: blue, slowly turns olive green (see Section V, 2a).

Salt of the ergocryptine:

Phosphate from 90% alcohol in hexagonal platelets, m.p. 198–200° (dec.).

Tartrate, from 10 fold the amount of methanol in thin rectangular platelets arranged in clusters m.p. 209° (dec.).

Ethanesulfonate, from alcohol when diluted with ether in prisms arranged in clusters, m.p. 204° (dec.).

Ergocryptinine
$C_{32}H_{41}O_5N_5$ (m.w. 575.7)

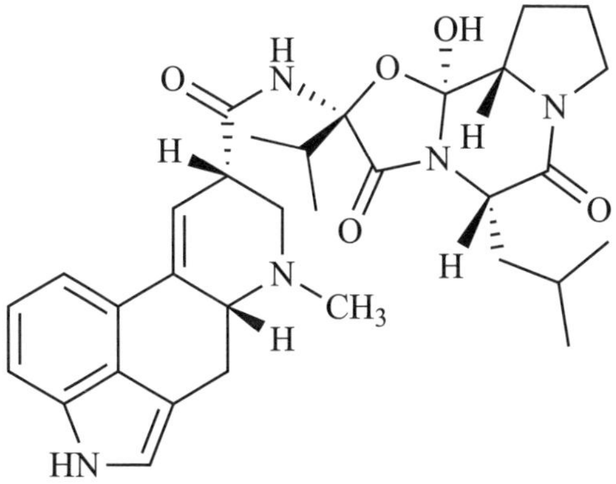

Ergocryptinine was prepared by rearranging ergocryptine in boiling methyl alcohol[71].

From ethanol, in which it dissolves in 20 parts at boiling, ergocryptinine crystallizes in long, thin prisms, m.p. 240–242° (dec.).

$pK^* = 4.9$ (*in 80% aq. methyl cellosolve*)
$[\alpha]_D^{20} = +399°$; $[\alpha]_{5461}^{20} = +501°$ (*c = 1.0 in chloroform*)
$[\alpha]_D^{20} = +479°$; $[\alpha]_{5461}^{20} = +598°$ (*c = 1.0 in pyridine*)

IR spectrum see Section V, 3b, Figure 15h.

Keller color reaction: blue, slowly turns olive green (see Section V, 2a).

Ergocornine

$C_{31}H_{39}O_5N_5$ (m.w. 561.7)

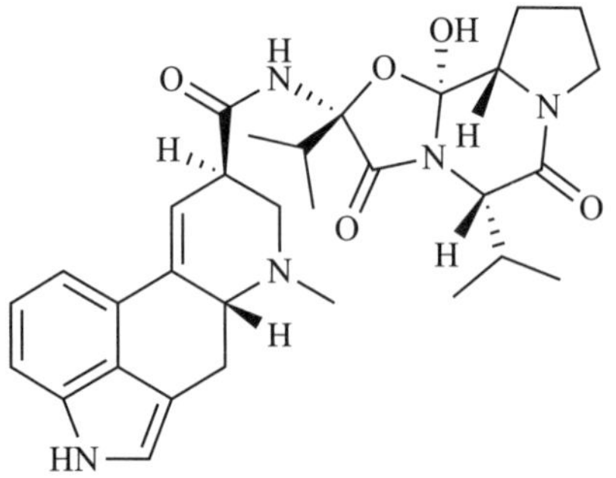

Ergocornine was first identified when the ergotoxine complex was separated with di-(p-toluyl)-L-tartaric acid, with which it forms the most soluble salt of the three components[71]. Once seed crystals of the pure alkaloid were available, it was possible, due to the poor solubility of ergocornine in methanol, to separate it in the form of the free base via fractional crystallization by dissolving ergotoxine preparations in a little methanol and inoculating them with ergocornine.

Ergocornine is also used medicinally as a dihydro derivative as a component of "Hydergine"®.

Ergocornine crystallizes from methanol, in which it is sparingly soluble, unlike the other alkaloids of the ergotoxine complex, in hard, strongly lustrous polyhedra, m.p. 182–184° (dec.).

It can also be easily recrystallized from ethyl alcohol or from acetone, in which it is moderately soluble. From 20 times the amount of boiling benzene, elongated, rectangular plates are obtained, which, like those of the other two alkaloids of the ergotoxine group, rearrange themselves on standing into solid prisms and polyhedra rich in surface area.

$pK^* = 5.5$ (*in* 80% *aq. methyl cellosolve*)
$[\alpha]_D^{20} = -186°$; $[\alpha]_{5461}^{20} = -224°$ ($c = 1.0$ *in chloroform*)
$[\alpha]_D^{20} = -111°$; $[\alpha]_{5461}^{20} = -129°$ ($c = 1.0$ *in pyridine*)

IR spectrum see Section V, 3b, Figure 15i.

Keller color reaction: blue, slowly turns olive green (see Section V, 2a).

Salts of ergocornine:

Hydrobromide, from acetone sharpened prisms, m.p. 225° (dec.).

Phosphate, from 90% alcohol in tufts of sharpened prisms, m.p. 190–195° (dec.).

Ethanesulfonate, from alcohol in long, thin prisms with triangular cross section, m.p. 209° (dec.)

Ergocorninine

$C_{31}H_{39}O_5N_5$ (m.w. 561.7)

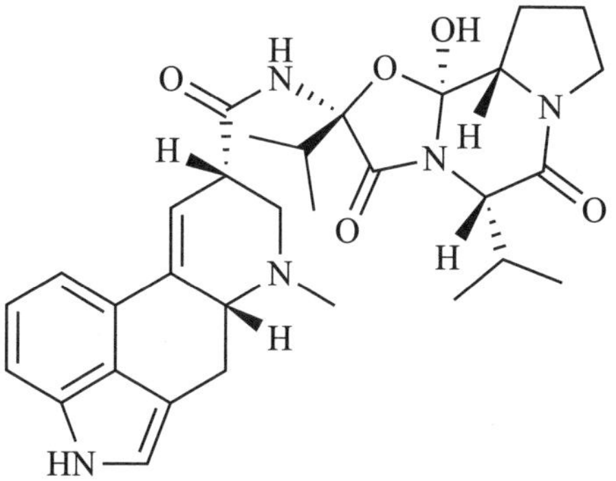

Ergocorninine was prepared by isomerization of ergocornine[71]. Since ergocorninine is, as an exception, more soluble in methyl alcohol than ergocornine, the otherwise common rearrangement by boiling in methyl alcohol fails for this pair of isomers. By letting ergocornine stand for a short time in alcoholic potassium hydroxide solution at room temperature, an equilibrium of the two isomers is established. From the resulting mixture, ergocorninine could be separated by fractional crystallization from ethyl acetate/ether.

Ergocorninine, unlike the other two dextrorotatory isomers of the ergotoxine complex, is readily soluble in methanol and in ethanol. It is very soluble in acetone and in chloroform. From 15 times the amount of alcohol, ergocorninine crystallizes in solid, pointed prisms, m.p. 228° (dec.).

$pK^* = 4.8$ (*in 80% aq. methyl cellosolve*)

$[\alpha]_D^{20} = +400°$; $[\alpha]_{5461}^{20} = +504°$ ($c = 1.0$ *in chloroform*)

$[\alpha]_D^{20} = +488°$; $[\alpha]_{5461}^{20} = +624°$ ($c = 1.0$ *in pyridine*)

IR spectrum see Section V, 3b, Figure 15k.

Keller color reaction: blue, slowly turns olive green (see Section V, 2a).

Ergostine

$C_{34}H_{37}O_5N_5$ (m.w. 595.7)

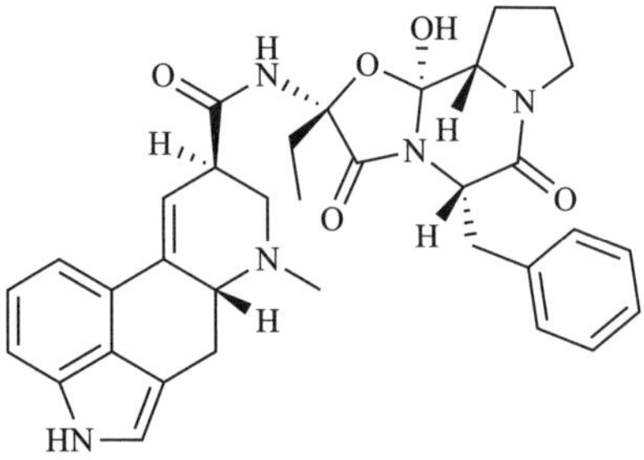

Ergostine is present in very small amounts in rye ergot as a concomitant alkaloid of ergotamine[80].

Ergostine crystallizes from ethyl acetate or from acetone in elongated prisms, m.p. 211–212° (dec.). The alkaloid is very slightly soluble in chloroform and in alcohol, moderately soluble in acetone, poorly soluble in ethyl acetate and in benzene.

$pK^* = 5.4$ (*in 80% aq. methyl cellosolve*)
$[\alpha]_D^{20} = -169°$; $[\alpha]_{5461}^{20} = -203°$ (*c = 1.0 in chloroform*)
$[\alpha]_D^{20} = -38°$; $[\alpha]_{5461}^{20} = -39°$ (*c = 1.0 in pyridine*)

In the Keller color reaction, a blue coloration is initially developed, which changes to blue-green after about 15 seconds.

IR spectrum see Section V, 3b, Figure 15 l.

Salts of ergostine:

Tartrate, from methanol in rhombic plates, m.p. 188–191° (dec.).

Hydrogen maleate, from methanol needles, m.p. 191–192° (dec.).

Ergostinine

$C_{34}H_{37}O_5N_5$ (m.w. 595.7)

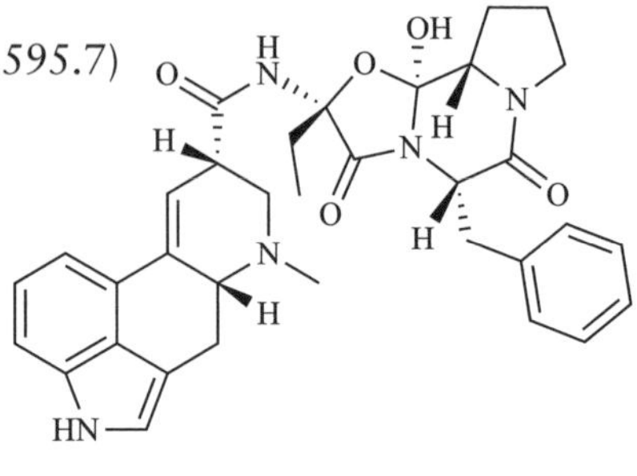

Ergostinine was prepared by rearrangement of ergostine[80].

Ergostinine crystallizes from 250 times the amount of boiling methanol or ethanol on cooling in long, pointed prisms. m.p. 215–216° (dec.). The alkaloid is readily soluble in chloroform and very slightly soluble in benzene.

$pK^* = 5.3$ (*in 80% aq. methyl cellosolve*)
$[\alpha]_D^{20} = +357°$; $[\alpha]_{5461}^{20} = +446°$ ($c = 1.0$ *in chloroform*)
$[\alpha]_D^{20} = +429°$; $[\alpha]_{5461}^{20} = +538°$ ($c = 1.0$ *in pyridine*)

In the Keller color reaction, a blue coloration is initially developed, which changes to blue-green after about 2-3 minutes.

IR spectrum see Section V, 3b, Figure 15 m.

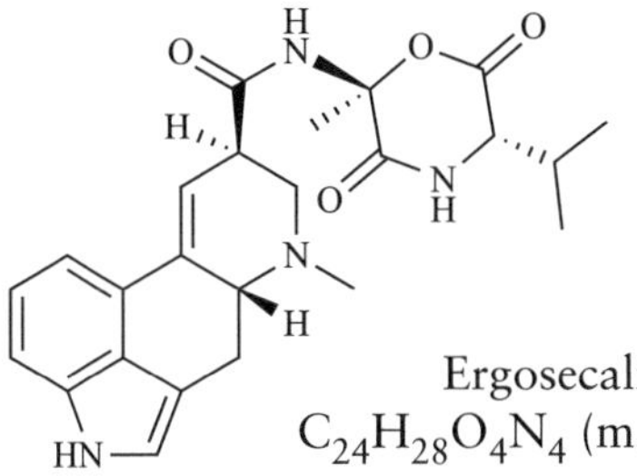

Ergosecaline
$C_{24}H_{28}O_4N_4$ (m.w. 436.5)

This alkaloid was detected by *M. Abe et al.*[75] in Spanish ergot by paper chromatography and was found to be identical to the rearrangement product of ergosecalinine. From that the above structure was deduced. It still requires further verification.

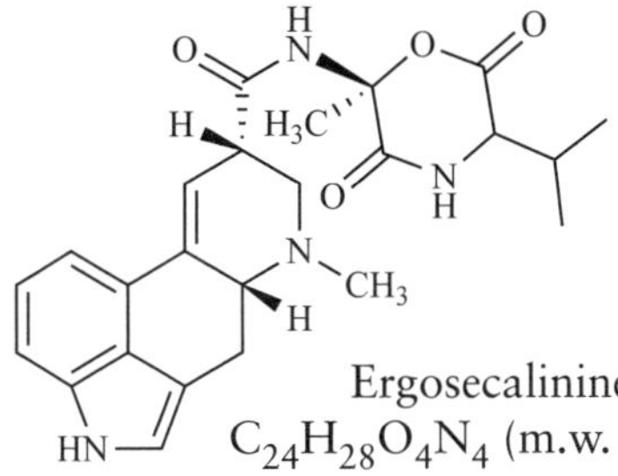

Ergosecalinine
$C_{24}H_{28}O_4N_4$ (m.w. 436.5)

Ergosecalinine was isolated by *M. Abe et al.*[75] in small amounts from Spanish ergot and from saprophytic cultures of this fungus. It could be reversibly rearranged into ergosecaline, the corresponding alkaloid of the lysergic acid series.

The above structure was derived solely from the hydrolysis products, namely lysergic acid, pyruvic acid and valine. It requires further verification.

From ethyl acetate prisms, m.p. 217° (dec.). Slightly soluble in methanol or acetone, moderately soluble in ethyl acetate and sparingly soluble in benzene or chloroform.

Ergobasine (Ergometrine, Ergonovine)

$C_{19}H_{23}O_2N_3$ (m.w. 325.4)

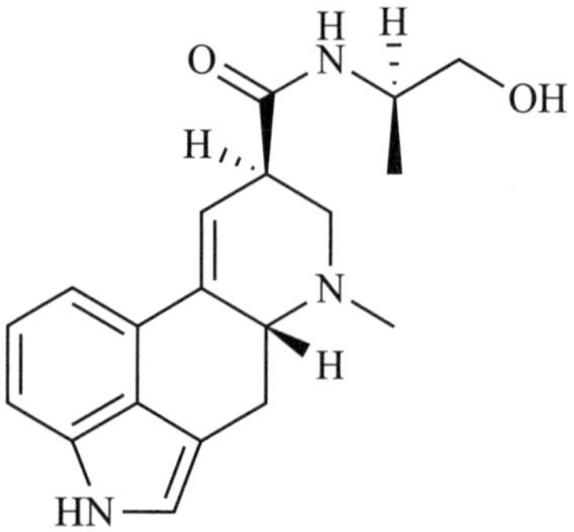

The observation of the English gynecologist C. *Moir* that aqueous extracts of ergot, which were free of ergotamine or ergotoxine alkaloids, had a strong oxytocic activity resulted in several laboratories working on the isolation of the water-soluble factor[108]. Three years after the finding published by C. *Moir* in 1932, a new, strongly uterus-active, relatively easily water-soluble alkaloid was described almost simultaneously by four working groups under four different names. *H. W. Dudley* and *C. Moir*[64] called the new alkaloid ergometrine, *A. Stoll* and *E. Burckhardt*[65, 66] ergobasine, *M. S. Kharasch* and *R. R. Legault*[67] ergotocine and *M. R. Thompson*[68] ergostetrine. The purest preparation, which first led to the correct gross formula, was in the hands of the chemists of the Basel laboratory. While the names ergobasine and ergometrine remained in use in Europe, ergonovine was introduced as the official name in the USA.

The content of ergobasine in ergot drugs of different origin, that is, in the different chemical strains, varies within wide limits. Whereas Portuguese commercial products, with a total alkaloid content of 0.20–0.25%, give yields of ergobasine of 0.01–0.02%, in Swiss ergot, with a total alkaloid content averaging 0.3% ergotamine, levels of 0.007–0.03% ergobasine were found.

Ergobasine crystallizes from 100 times the amount of boiling ethyl acetate on cooling in bulky polyhedra, often displaying tetrahedral form, from 400 times the amount of benzene in soft needles, from 10 parts of methyl ethyl ketone in obtuse prisms melting at 162° (corr.) with decomposition. From chloroform, in which the alkaloid is very sparingly soluble, it separates with 1 mol of crystallization solvent. The chloroform compound is suitable for the separation of ergobasine from impurities which are easily soluble in chloroform. The alkaloid is moderately soluble in water.

Dimorphism was observed during crystallization from acetone[107]. In addition to the form melting at 162°, long needles of m.p. 212° (dec.) were obtained from this solvent. The lower melting modification changes into the more stable, higher melting form during storage.

$$pK^* = 6.0 \; (in\; 80\%\; aq.\; methyl\; cellosolve)$$
$$[\alpha]_D^{20} = +90° \; (c = 0.25\; in\; water)$$
$$[\alpha]_D^{20} = +41°; \; [\alpha]_{5461}^{20} = +60° \; (c = 1.0\; in\; alcohol)$$

UV spectrum is the same as that of lysergic acid and isolysergic acid.

IR spectrum see Section V, 3b, Figure 15 n.

Keller color reaction: blue (see Section V, 2a).

Salts of ergobasine:

Hydrochloride, $C_{19}H_{23}O_2N_3 \cdot HCl$, m.p. 245–256° (corr.) (dec.Table 1).

Tartrate, $(C_{19}H_{23}O_2N_3)_2 \cdot C_4H_6O_6 \cdot 2\;CH_3OH$. This easily water-soluble salt crystallizes from methanol with crystal solvent, m.p. indistinct between 130–160°.

$$[\alpha]_D^{20} = +63° \; (c = 0.25\; in\; water)$$

Hydrogen maleate, $C_{19}H_{23}O_2N_3 \cdot C_4H_4O_4$. From methanol in white needles of m.p. 188–190° (dec.).

$$[\alpha]_D^{20} = +53° \, (c = 0.5 \text{ in water})$$

This salt has been included in most pharmacopoeias, as ergonovine maleate in the US Pharmacopoeia USP XVI, as ergometrine maleate in the British Pharmacopoeia BP 58, and as ergobasine maleate in the Austrian Pharmacopoeia ÖAB 9.

Ergobasinine (ergometrinine)
$C_{19}H_{23}O_2N_3$ (m.w. 325.4)

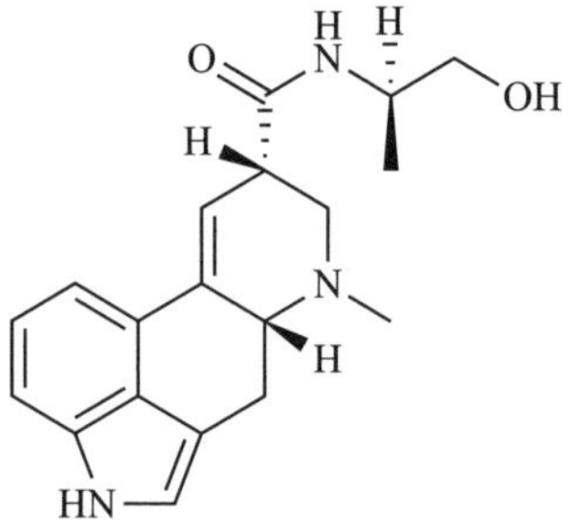

Ergobasinine is present in small amounts in ergot along with ergobasine[110].

It can be prepared from ergobasine by rearrangement in alkaline environment[109].

Ergobasine crystallizes readily from acetone in massive, clear prisms that melt at 196° (corr.) with decomposition.

$$[\alpha]_D^{20} = +414°; [\alpha]_{5461}^{20} = +520° \, (c = 1.0 \text{ in chloroform})$$
$$pK^* = 6.2 \, (\text{in } 80\% \text{ aq. methyl cellosolve})$$

UV spectrum: same as that of lysergic acid and isolysergic acid.

IR spectrum: see Section V, 3b, Figure 15 o.

Keller color reaction: blue (see Section V, 2a)

Salts of ergobasinine[109]:

Hydochloride, $C_{19}H_{23}O_2N_3 \cdot HCl$, from aqueous acetone needles, m.p. 175–180° (dec.).

Perchlorate, $C_{19}H_{23}O_2N_3 \cdot HClO_4$, needles, m.p. 225° (dec.).

Nitrate, $C_{19}H_{23}O_2N_3 \cdot HNO_3$, blunt prisms, m.p. 235° (dec.).

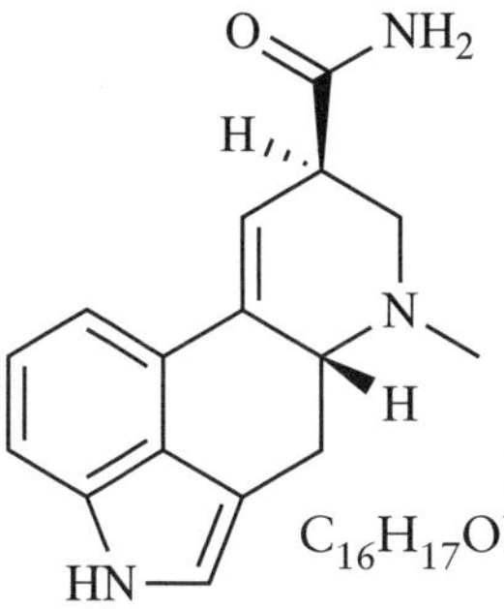

Ergine
$C_{16}H_{17}ON_3$ (m.w. 267.3)

Lysergic acid amide and isolysergic acid amide, which were for a long time known only as hydrolysis products of ergot alkaloids[81, 88,111], have recently been isolated as main components of the alkaloid mixture from the ergot of the wild grass *Paspalum distichum* L.[20], and are therefore also to be regarded as true ergot alkaloids. The term ergine, which was originally used for the first characteristic hydrolysis product of ergot alkaloids before it was identified as isolysergic acid amide[81], is now correctly applied to lysergic acid amide. Isolysergic acid amide is therefore to be called isoergine, or, since ergine and isoergine have now been found to be genuine ergot alkaloids, according to the nomenclature adopted here, erginine.

Recently, ergine and erginine have been isolated as major components, among other alkaloids of the clavine group, from the seeds of *Rivea corymbosa* (L.) Hall. f. and *Ipomoea tricolor* Cav. of the family *Convolvulaceae*[78, 112]. The seeds of these two morning glory species were used centuries ago by the Aztecs under the name "Ololiuqui" as a magic drug, and are still used today in the remote mountains of southern Mexico for magical purposes[113]. The brown seeds of *Rivea corymbosa* are called "badoh" in the Zapotec language, the black seeds of *Ipomoea tricolor* "badoh negro." "Ololiuqui" is, along with the well-known mescaline drug "Peyotl" (aka peyote, *Lophophora williamsii* [Lem.] J.M. Coult.), a type of cactus, and the sacred mushroom "Teonanacatl" (*Psilocybe* spp.)[274] one of the three most important magic drugs used by the Aztecs and related peoples in their religious ceremonies and medical practices.

The occurrence of lysergic acid alkaloids, previously found only in lower fungi of the genus *Claviceps* and more recently also in the genera *Aspergillus* and *Rhizopus*[79], in the plant family *Convolvulaceae* is a quite unexpected phytochemical finding.

Ergine crystallizes from methanol in prisms which melt at 242° (corr.) with decomposition.

$$[\alpha]_D^{20} = 0° (\pm 2°); [\alpha]_{5461}^{20} = + 15° (c = 0.5 \text{ in pyridine})$$
$$pK^* = 5.9 \text{ (in 80\% aq. methyl cellosolve)}$$

IR spectrum see Section V, 3b, Figure 15p.

Ergine methanesulfonate, $C_{16}H_{17}ON_3 \cdot CH_3SO_3H$. This readily water-soluble salt crystallizes from the concentrated solution in methanol when diluted with acetone in elongated prisms of m.p. 232° (dec.).

Erginine
$C_{15}H_{17}ON_3$ (m.w. 267.3)

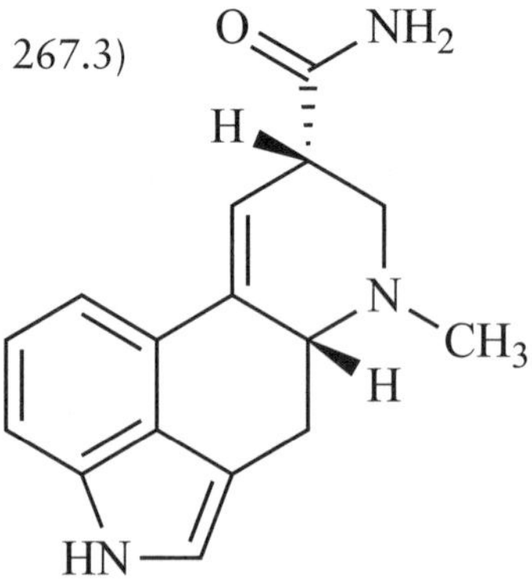

Erginine crystallizes from solution in 200 parts of boiling methanol on cooling in elongated prisms containing 1 mol of crystal solvent, m.p. 132–134°.

$$[\alpha]_D^{20} = + 480°; [\alpha]_{5461}^{20} = + 608° (c = 0.5 \text{ in pyridine})$$
$$pK^* = 6.1 \text{ (in 80\% aq. methyl cellosolve)}$$

IR spectrum see Section V, 3b, Figure 15q.

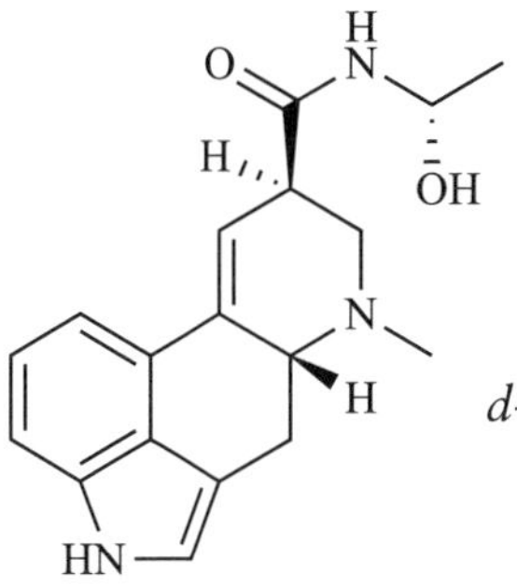

d-Lysergic acid methyl carbinolamide
$C_{18}H_{21}O_2N_3$ (m.w. 311.4)

This alkaloid, isolated from saprophytic cultures of *Claviceps paspali* Stevens and Hal alongside ergine and erginine[20], readily decomposes in aqueous solution into ergine and acetaldehyde. It does not possess any pharmacological properties that might suggest its use in medicine[368].

From chloroform long prisms, free of crystalline solvent, m.p. 135° (dec.).

$$[\alpha]_D^{20} = +\ 29° \ (c = 1.0 \ in \ dimethlyformamide)$$

IR spectrum: maxima at 242 and 312 nm.

2. Lysergic acid, isolysergic acid, and dihydrolysergic acids

Before going into the structural determination of the individual ergot alkaloids, it is useful to first discuss the chemistry of the common scaffold of all these alkaloids, lysergic acid, as well as their isomers and closest derivatives.

a) Extraction and properties of lysergic acid and isolysergic acid as well as their methyl esters and hydrazides

The ergot alkaloids of main group A (cf. Section C, II) give during vigorous alkaline hydrolysis (boiling for one hour in 7% aqueous sodium hydroxide) a mixture of lysergic acid and isolysergic acid as a characteristic cleavage product[82, 83, 114].

If the hydrolytic cleavage of ergot alkaloids is carried out more gently, by boiling with only a small excess of alcoholic lye, then the lysergic acid part precipitates in the form of the amide, and instead of lysergic acid and isolysergic acid, a mixture of lysergic acid amide and isolysergic acid amide is obtained[81, 88, 111]. Both of these amides were later found as genuine alkaloids in *Paspalum* ergot[20] and in the seeds of certain *Convolvulaceae*[112], and described as ergine, resp. erginine (see Section C, III/1).

Lysergic acid and isolysergic acid occur in nature not only in the form of alkaloid derivatives, but have also been found in very small quantities as such in saprophytic cultures of the ergot fungus[115].

Lysergic acid, $C_{16}H_{16}O_2N_2$, crystallizes from water, in which it dissolves at boiling in 150 parts, in thin, elongated hexagonal platelets containing 1 molecule of crystallization water. The acid is readily soluble in pyridine, moderately to sparingly soluble in methanol and ethanol, and very sparingly soluble in most other organic solvents. The specific rotation value is $[\alpha]_D^{20} = + 10°\,(c = 1.0\ in\ pyridine)$.

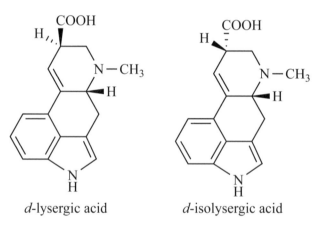

d-lysergic acid *d*-isolysergic acid

Higher values are also given in the technical literature, which is due to the fact that it is very difficult to free the lysergic acid from small amounts of the strongly dextrorotatory isolysergic acid.

The natural, dextrorotatory lysergic acid was named *d*-lysergic acid and its isoform *d*-isolsergic acid[116].

Their optical antipodes, *l*-lysergic acid and *l*-isolysergic acid, do not occur in nature and can only be accessed by racemization of the optically active lysergic acid[116] or by total synthesis of the racemates[96, 117]. For simplicity, the prefix *d* is usually omitted where natural lysergic acid is referred to. Lysergic acid reacts amphoterically. It is easily soluble in aqueous alkali as well as in acids under salt formation. It can be precipitated from the aqueous solution of the alkali salt by introducing carbonic acid. The easily soluble hydrochloride of lysergic acid, $C_{16}H_{16}O_2N_2 \cdot HCl$, m.p. 208–210° (dec.), crystallizes from dilute hydrochloric acid. The sulphate $(C_{16}H_{16}O_2N_2)_2 \cdot H_2SO_4$ is somewhat more difficult to dissolve and crystallizes in platelets which melt at 220° (dec.).

Through the exposure to acid or alkali, but also by boiling in water, the lysergic acid is partly converted into isolysergic acid[88]. An equilibrium is established in the solutions. The isomerization is reversible. Isolysergic acid is slightly more soluble in water than lysergic acid and crystallizes from it with 2 H_2O, $[\alpha]_D^{20} = +281°$; $[\alpha]_{5461}^{20} = +368°$ ($c = 1,0$ *in pyridine*). Isolysergic acid is slightly more basic than lysergic acid. *pK* (*HCl in* 80% *aq. methylcellosolve*) = 4.4 *for lysergic acid* and *4.0 for isolysergic acid;* *pK* (*tetramethylammonium hydroxide in* 80% *aq. methylcellosolve*) = 7.9 *for lysergic acid and 8.8 for isolysergic acid.*

The dextrorotatory lysergic acid could be racemized by heating for four hours with barium hydroxide at 150°[88].

Racemization of the lysergic acid molecule is particularly easy when the ergot alkaloids or an ester of lysergic acid are reacted with hydrazine or hydrazine hydrate on heating to obtain the lysergic acid hydrazide[116]. Partial isomerization occurs simultaneously, so that the less soluble racemic isolysergic acid hydrazide is formed in this treatment. The racemate was either resolved by fractional crystallization after conversion to the azide and linkage with *l*-norephedrine, or by salt formation with di-(*p*-toluyl)-L- or -D-tartaric acid into the optical antipodes, into the *d*- and *l*-isolysergic acid hydrazides[117]. The optically active lysergic acid and isolysergic acid hydrazides are valuable starting materials for the synthesis of lysergic acid derivatives.

It is possible to avoid racemization of the lysergic acid residue by reacting the lysergic acid derivatives with hydrazine in the presence of at least one equivalent of protons. To suppress racemization, it is sufficient to use ergotamine hydrochloride instead of the free ergotamine base for the cleavage, or one can replace part of the hydrazine with hydrazine dihydrochloride[118].

Table 2

Properties of the isomeric lysergic acids and their methyl esters

Substance	M.p. (dec.)	$[\alpha]_D^{20}$	Typical crystallization
Acid: $C_{16}H_{16}O_2N_2$ (268.3)			
d-Lysergic acid	238°	+ 32° (pyridine)	From water, elongated hexagonal leaflets with 1 mol. H_2O.
l-Lysergic acid	238°	− 2° (pyridine)	
d-Isolysergic acid	218°	− 281° (pyridine)	From water, in elongated 3-sided leaflets with 2 mol. H_2O.
rac. Lysergic acid	240-50°	—	From water, in rectangular leaflets with 1 mol. H_2O.
rac. Isolysergic acid	240-50°	—	From water, in rhombic leaflets with 1 mol. H_2O.
Methylester: $C_{17}H_{18}O_2N_2$ (282.3)			
d-Lysergic acid methylester	170°	+ 82° ($CHCl_3$)	From benzene, in leaflets
d-Isolysergic acid methylester	174°	+ 179° ($CHCl_3$)	From benzene, in thin rods

Table 3

Properties of the lysergic and isolysergic acid hydrazides
$C_{16}H_{18}ON_4$ (282.3)

Substance	M.p. (dec.)	$[\alpha]_D^{20}$ (in pyridine)	Solubility and typical crystallization
d-Lysergic acid hydrazide	218°	+ 11°	From 50 times the amount of methanol, in long, thin prisms.
l-Lysergic acid hydrazide	218°	− 11°	
d-Isolysergic acid hydrazide	204°	+ 452°	Easily soluble in methyl and ethyl alcohol; from it in massive prisms.
l-Isolysergic acid hydrazide	204°	− 454°	
rac. Lysergic acid hydrazide	220°	—	From 100 parts of hot alcohol, in long needles.
rac. Isolysergic acid hydrazide	240°	—	From 300 parts of hot alcohol, in hexagonal plates.

By the action of diazomethane, the methyl esters of lysergic acid and isolysergic acid can be prepared[88].

Tables 2 and 3 summarize the most important data for the isomeric lysergic acids and the derivatives discussed.

Lysergic acid and isolysergic acid are easily decomposable. They are especially sensitive to oxidative influences, to strong acids and to light. Most of the derivatives of lysergic acid and the ergot alkaloids also show this great decomposability, which explains why, compared to other alkaloid drugs, it was only relatively recently that it was possible to produce the alkaloids from ergot in pure crystallized form.

Lysergic acid and its derivatives give typical color reactions (*Keller* reaction, *van Urk-Smith* reaction). These color tests, which are important for the detection of the ergot alkaloids, are discussed in more detail in the chapter on analytics (Section V, 2).

b) Preparation and properties of isomeric dihydrolysergic acids, their methyl esters, hydrazides, and amides

The catalytic hydrogenation of lysergic acid and isolysergic acid or their derivatives, e.g., the natural ergot alkaloids, with platinum catalyst in glacial acetic acid produces the dihydro derivatives of lysergic acid and isolysergic acid[119-121]. While a unitary dihydro compound is formed in the lysergic acid series, two stereoisomeric dihydro compounds are obtained from the isolysergic acid and its derivatives, which were named dihydroisolysergic acid-(I) and dihydroisolysergic acid-(II), resp. dihydroergotaminine-(I) and dihydroergotaminine-(II). The corresponding isomeric dihydrolysergic acids can also be obtained via the alkaline hydrolysis of the dihydro alkaloids. These can be converted into the methyl esters either with diazomethane or with methanol/HCl. The isomeric dihydrolysergic acid hydrazides are formed from the methyl esters or the dihydro alkaloids when boiled with hydrazine. These have been used as starting materials for numerous derivatives. The isomeric dihydrolysergic acid amides are thus obtained via the dihydrolysergic acid azides with ammonia.

The hydrogen addition has stabilized the molecule, because neither isomerization nor racemization takes place during cleavage with hydrazine, whereas the non-hydrogenated alkaloids lead to the racemic isolysergic acid hydrazide when exposed to hydrazine.

Table 4 summarizes the most important properties of the isomeric dihydrolysergic acids and some of their derivatives.

Table 4
Properties of the isomers of dihydrolysergic acids,
their hydrazides, methyl esters and amides

	Dihydrolysergic acid	Dihydroisolysergic acid-(I)	Dihydroisolysergic acid-(II)
Acid: $C_{16}H_{18}O_2N_2$ (270.3) Melting point $[\alpha]_D^{20}$ (in pyridine) Crystallization from water	318° (Block[1]) − 122° hexagonal leaflets	280° (Block[1]) − 86° irregular leaflets	310° (Block[1]) + 17° massive polyhedra
Hydrazide: $C_{16}H_{20}ON_4$ (284.3) Melting point $[\alpha]_D^{20}$ (in pyridine) Crystallization from methanol	247° − 123° needles	227° − 23° needles	260° + 56° needles
Methyl ester: $C_{17}H_{20}O_2N_2$ (284.3) Melting point $[\alpha]_D^{20}$ (in pyridine) Crystallization from methanol	187° − 96° long prisms	190° − 82° long prisms	amorphous
Amide: $C_{16}H_{19}ON_3$ (269.3) Melting point $[\alpha]_D^{20}$ (in pyridine) Crystallization from methanol	276° (Block[1]) − 131° prisms and plates	275° (Block[1]) 0° 4- or 6-sided plates	307° (Block[1]) + 17° prisms

[1])Melting points followed by "Block" are uncorrected; all others are corrected but, owing to the proximity of melting and decomposition temperatures, should not be viewed as absolutely accurate, but relative to closely allied structures[121].

c) Structural elucidation of lysergic acid

On the basis of relatively few degradation products, W. A. Jacobs and L. C. Craig succeeded with astonishing intuition in deriving the ring structure of lysergic acid. By the oxidation of ergot alkaloids and of lysergic acid, they obtained a quinoline betaine tricarboxylic acid,

which, though only later, could be assigned Structure I [183, 122, 123].
The alkali decomposition of dihydrolysergic acid yielded methylamine,
propionic acid, 1-methyl-5-amino-naphthalene (Structure II), and
3,4-dimethylindole (Structure III)[124, 125].

From these cleavage pieces, the aforementioned authors concluded
that a novel tetracyclic ring system (Structure IV) must underlie the lysergic acid.

Based on hydrogenation experiments and the comparison of the UV
spectra of lysergic acid and dihydro derivatives, it could be concluded that
the double bond of lysergic acid located outside the indole system must
be in conjugation with the unsaturated system of the indole[126]. From pK
measurements of lysergic acid, isolysergic acid and dihydro derivatives[127]
and from the determination of a ß-amino carboxylic acid cleavage (for-
mulations p. 49) on dihydrolysergic acid[90], the position of the carboxyl
group could also be determined, so that the formulas V and VI were
proposed for lysergic acid and isolysergic acid, respectively.

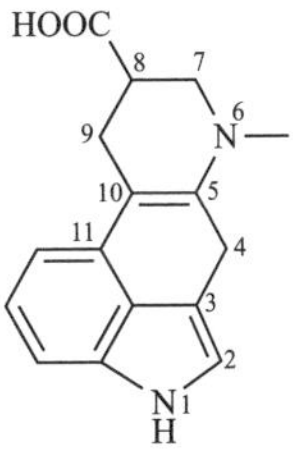

which, though only later, could be assigned Structure I [183, 122, 123].

V 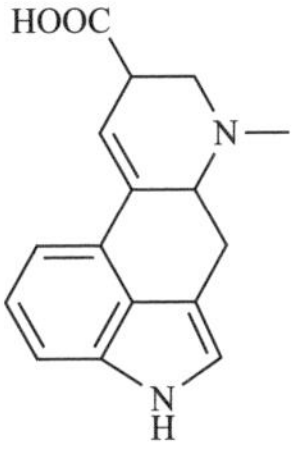   VI 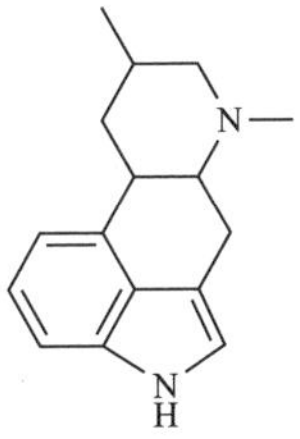   VII

The ring structure of lysergic acid as represented by these formulae
and the position of the carboxyl group in position 8 could be confirmed
by the synthesis.

This was a long and difficult task, because the tetracyclic ring skeleton of lysergic acid was new and the methods for its synthesis had to
be developed first. *W. A. Jacobs* and *R. G. Gould Jr.* proposed, for the
tetracyclic ring system IV saturated in the C and D rings, the name "ergoline." In 1937, they succeeded in synthesizing this basic structure of the

ergot alkaloids[101]. This synthesis is discussed in more detail in Section III, 2f). Later, the same authors also produced 6-methylergoline[128]. However, the assignment of a totally synthetic compound with a breakdown product did not succeed until racemic lysergic acid could be converted via its dihydro derivative to the degradation product VII, which proved to be identical with synthetic racemic 6,8-dimethylergoline[129, 130]. This proved the ring system of lysergic acid and the position of the carboxyl group at C8.

The novel feature of ergoline is the ring closure at position 4 of indole (= position 11 of ergoline), which distinguishes the ergot alkaloids containing this ring system from all previously known indole alkaloids and other naturally occurring indole derivatives, in which further rings are always connected to positions 1, 2, 3 or 7.

Further structural features of lysergic acid are the ring D, which is present as an *N*-methylated, tetrahydrogenated nicotinic acid ring, and the free α-position of the indole system, which is responsible for the positive outcome of the *van Urk-Smith* color reaction typical of ergot alkaloids with *p*-dimethylaminobenzaldehyde and acid.

d) The lysergic acid-isolysergic acid isomerism[92]

However, according to the investigations discussed in the previous section, the position of the double bond outside the indole system, which is in conjugation with it, was not yet certain, and its different position according to formulae V and VI was initially held responsible for the isomerism between lysergic acid and isolysergic acid. The lysergic acid and isolysergic acid would have been structural isomers based on this formulation.

This last uncertain point in the structural formulas of lysergic acid and isolysergic acid could be clarified when it was shown that upon removing the center of asymmetry at C8 from lysergic acid and isolysergic acid, an identical derivative is formed, which is still optically active. The center of asymmetry at C8 could be removed by briefly heating the lysergic acid, the isolysergic acid or the isomeric dihydrolysergic acid with acetic anhydride. This treatment opens the ring D in the manner of a ß-amino carboxylic acid cleavage between the nitrogen atom and C7, after which lactam formation occurs between the resulting secondary amino group and the carboxyl group.

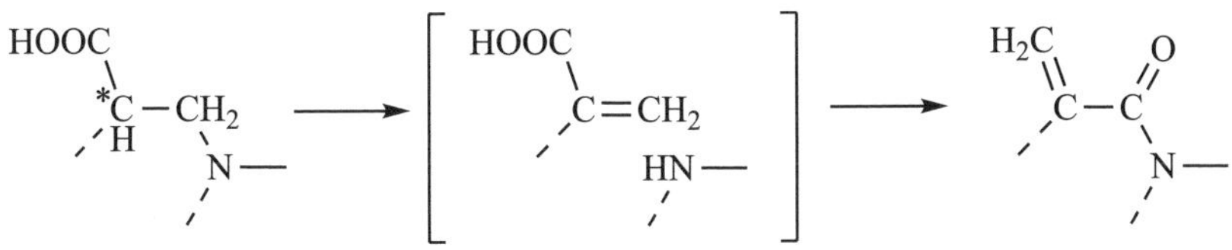

$$
\begin{array}{ccc}
\underset{\underset{,'H}{\overset{*}{C}}-CH_2}{\overset{HOOC}{|}} \ \underset{\overset{|}{N-}}{} & \longrightarrow & \left[\ \underset{\underset{HN-}{}}{\overset{HOOC}{|}}\ C=CH_2 \ \right] & \longrightarrow & \underset{\underset{N-}{}}{\overset{H_2C}{\backslash\backslash}}\ C-C\overset{O}{\overset{//}{}}
\end{array}
$$

The double bond that occurs during this cleavage removes the center of asymmetry at C8 and, most importantly, the same optically active lactam, $C_{16}H_{14}ON_2$, is formed from the lysergic acid and from the isolysergic acid in equally good yield. It crystallizes from methanol in massive plates and prisms and has a specific rotation $[\alpha]_D^{20} = -35°$ ($c = 0.2$ in alcohol). When heated in a capillary tube, the lactam turns dark from 180° and decomposes above 300° without actually melting.

From these findings could be concluded:

a) Lysergic acid and isolysergic acid differ only in the steric arrangement at the asymmetric carbon atom 8. The rest of the molecules is identical in both isomers, in particular also the arrangement of the carbon double bonds, which is consistent with the identity of the UV absorption spectra of lysergic acid and isolysergic acid (see Figure 8).

b) The reducible double bond conjugated to the indole system can only be in the 4–5 or 9–10 position. The position in 5–10 must be ruled out, because under this assumption a center of asymmetry would only be possible at C8. If the asymmetry at C8 were removed, an optically inactive lactam would have to be formed, but this is not the case.

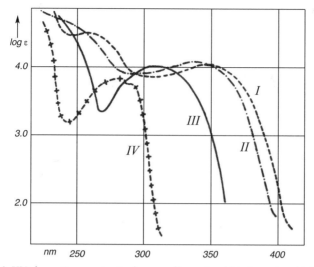

Figure 8. UV absorption spectra: I = lactam of lysergic acid-isolysergic acid; II = decarbox-ylation product of lysergic acid-isolysergic acid; III = lysergic acid-isolysergic acid (in the form of ergobasine-ergobasinine); IV = dihydrolysergic acid hydrochloride. Alcohol served as solvent for I to III, alcohol + HCl (0.01 N) for IV. I and II contain a new, conjugated C=C double bond, formed by the opening of ring D.

The decision between the two possible positions of the reducible dou-ble bond (Δ^{4-5} or Δ^{9-10}) can be made from the comparison of the UV absorption spectrum of the lactam with that of the acids. The strong shift of the UV absorption of the lactam towards the long wavelength region (cf. Figure 8) indicates that the new double bond that has entered the lactam in the 7–8 position is in conjugation with the already existing unsaturated system of the lysergic acid/isolysergic acid. It can be con-cluded that the reducible double bond of the lysergic acid and isolysergic acid occurs at the position 9–10, because in position 4–5 one misses the conjugation with the new double bond of the lactam in the 7–8 position.

From these experimental results, it could be concluded that lysergic acid and isolysergic acid are diastereomers and not structural isomers. They differ in the arrangement of the substituents at the asymmetric carbon atom 8. In both isomers, the easily hydrogenated carbon double bond is in position 9–10.

In addition to the described ß-amino carboxylic acid cleavage, the center of asymmetry at C8 could also be removed by the *Hofmann* degradation of a suitable lysergic acid derivative, and by decarboxylation of lysergic acid, whereby identical degradation products were obtained in both cases from the lysergic acid and the isolysergic acid.

The starting materials for the *Hofmann* degradation were 6-methyl-8-acetylamino-ergolene and 6-methyl-8-acetylamino-isoergolene[131]. The iodomethylates of the two isomeric ergolenes represent exquisitely crystallizing compounds. Upon liberating the quaternary bases with silver oxide, a small percentage spontaneously decomposed to the des-base in both series. The des-bases from the lysergic acid and isolysergic acid series were identical. Thus, ring opening must have occurred between N6 and C7.

6-Methyl-8-acetylamino-ergolene iodomethylate

6-Methyl-8-acetylamino-isoergolene iodomethylate

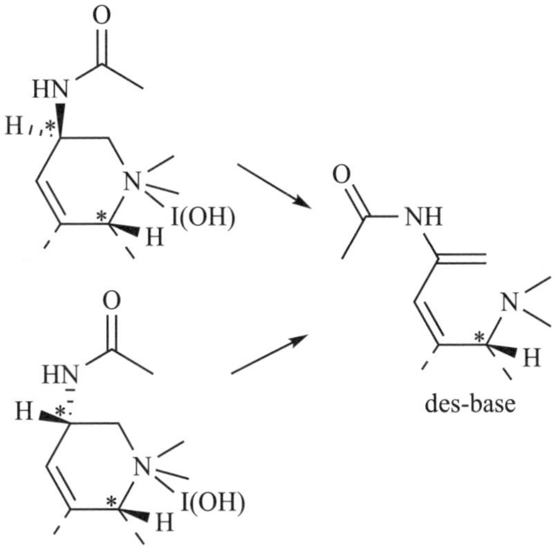

des-base

The des-base, which crystallizes well when freshly prepared, but soon darkens in the light, is conspicuous for its high specific rotation.

The simplest way to eliminate the asymmetry at C8 is the decarboxylation of the lysergic acid. This reaction, the use of which is severely limited in practice by the high decomposability of the lysergic acid molecules, could be carried out by introducing the acids in finely powdered form into boiling Dowtherm (b.p. 251°) and cooling immediately. The high specific levorotation of the decarboxylation product $[\alpha]_D^{20} = -325°$ (*in chloroform*) and its UV absorption spectrum (see

Figure 8) indicate that not only decarboxylation has taken place but that the pyrogenic conversion has reopened the ring D again between N6 and C7, which is evidently a weak site, with an introduction of a double bond conjugated to the existing system.

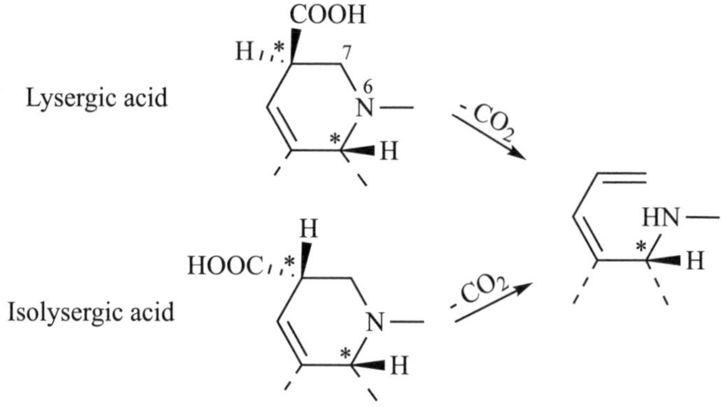

The formulation of lysergic acid and isolysergic acid as optical isomers at C8 also corresponds to the number of racemates observed. Since in lysergic acid, as in isolysergic acid, two centers of asymmetry are present when the double bond is fixed in the 9–10 position, at C5 and C8, two diastereomeric racemates must be formed when the molecules are racemized. Indeed, these two racemates are known and have been described in the form of racemic lysergic acid and racemic isolysergic acid[116] as well as in the form of racemic lysergic acid hydrazide and racemic isolysergic acid hydrazide[117].

The observations made in the hydrogenation of lysergic and isolysergic acid derivatives can also be easily formulated on the basis of the optical isomerism between lysergic acid and isolysergic acid. The saturation of the double bond in the 9–10 position in lysergic acid and in isolysergic acid with hydrogen leads to the formation of a new asymmetry center at C10, thus providing the possibility for two stereoisomeric dihydro derivatives. Whether both stereoisomeric dihydro derivatives are then formed actually in detectable quantities depends on the steric proportions and the hydrogenation conditions applied.

During the hydrogenation of lysergic acid or its derivatives, the formation of only one dihydro derivative was observed. The isolysergic acid alkaloids, on the other hand, give two stereoisomeric dihydro derivatives (I and II), whose mutual quantity ratio depends on the catalyst used[121].

Furthermore, it was observed that dihydroisolysergic acid-(I) can be irreversibly rearranged into dihydrolysergic acid under certain conditions. Thus, these two dihydro acids differ only in the steric arrangement of the carboxyl group. At C10, which has become asymmetric through hydrogenation, they have the same configuration. The finding that, in the case of the dihydro acids, only this one-sided rearrangement of the dihydroisolysergic acid-(I) into dihydrolysergic acid is possible is probably explained by the fact that dihydrolysergic acid is a particularly stable and therefore preferred steric arrangement.

It is consistent with this view that the hydrogenation of lysergic acid and its derivatives produces only this one isomer.

The more difficult isomerization in the dihydro acids compared to the spontaneous, reversible steric rearrangement of the carboxyl group of the natural lysergic acid also indicates the importance of the double bond (Δ^{9-10}), which the dihydro derivatives lack, in the rearrangement mechanism.

Lactonization of dihydroisolysergic acid-(I) with acetic anhydride produces the same lactam as from dihydrolysergic acid. This also shows that these two acids differ only in the steric arrangement of the carboxyl group. The lactam from dihydroisolysergic acid-(II) differs from the anhydrization product of dihydrolysergic acid and dihydroisolysergic acid-(I) by its mirror-image configuration at C10.

All these observations could be drawn out according to Formula Scheme 1 below (see p. 55).

In these diagrams, the steric arrangement at C10 is arbitrary for the time being. It should only be shown that the dihydrolysergic acid and the dihydroisolysergic acid-(I) have a matching configuration at C10 and that the dihydroisolysergic acid-(II) has a mirror-image arrangement at this center of asymmetry. The stereochemistry of the lysergic and dihydroly-sergic acids will be discussed in detail in the next section.

The experimental data discussed and further findings yet to be discussed also allow conclusions to be drawn about the mechanism of lysergic acid–isolysergic acid rearrangement. Lysergic acid and isolysergic acid as well as their natural derivatives, the ergot alkaloids, readily rearrange into each other in hydroxyl-containing solvents. Isomerization is accelerated both by acids and especially by alkalis. On the other hand, all derivatives of lysergic acid and isolysergic acid presented up to now, in which the carboxyl group is replaced by another residue, such as 6-methyl-8-amino-ergolene, 6-methyl-8-amino-isoergolene, as well as the 8-acetylamino derivatives of these two compounds[131], and also 6-methyl-ergolen-8-yl-carbamic acid ester and 6-methyl-isoergolen-8-yl-carbamic acid ester, are no longer isomerizable[132]. As already mentioned, dihydrolysergic acid and its derivatives can no longer be isomerized, whereas dihydroisolysergic acid-(I) can be unidirectionally converted into dihydrolysergic acid under activating conditions.

Accordingly, the carboxyl group in particular, but then also the reducible double bond, influences the isomerization process or is directly involved in it. From this point of view, the epimerization process, which must take place on the basis of the preceding considerations, could be plausibly explained by the following: the presence of a double bond in the 9,10-position favors the enolization of the carbonyl of the carboxyl group, because together with it, a continuously conjugated system of double bonds from the enol double bond to the indole system is formed. However, since the enol form at C8 is symmetrical, an equilibrium of the mirror-image compounds at carbon atom 8, that is, of the lysergic acid and isolysergic acid forms, is established thereby, as illustrated by the formulae on the facing page. It is understandable that the nature of the substituent R on the carboxyl group also has an influence on the rate of isomerization and on the position of the established equilibrium, which corresponds to experiment.

Formula Scheme 1

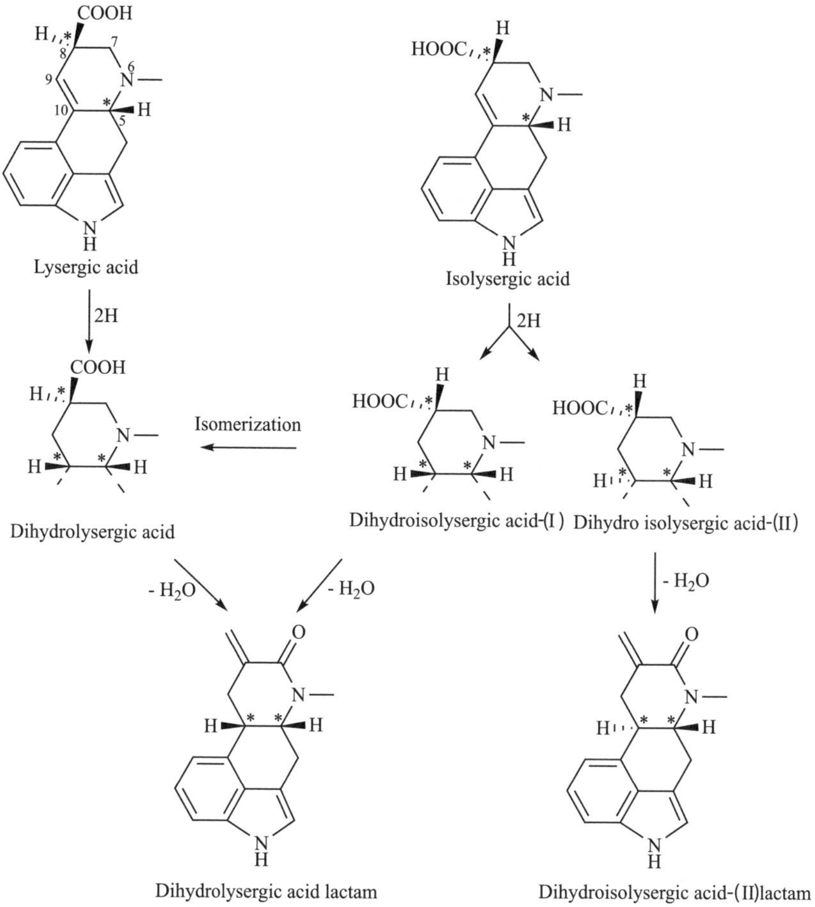
Lysergic acid
Isolysergic acid
2H
2H
Isomerization
Dihydrolysergic acid
Dihydroisolysergic acid-(I)
Dihydro isolysergic acid-(II)
- H_2O
- H_2O
- H_2O
Dihydrolysergic acid lactam
Dihydroisolysergic acid-(II)lactam

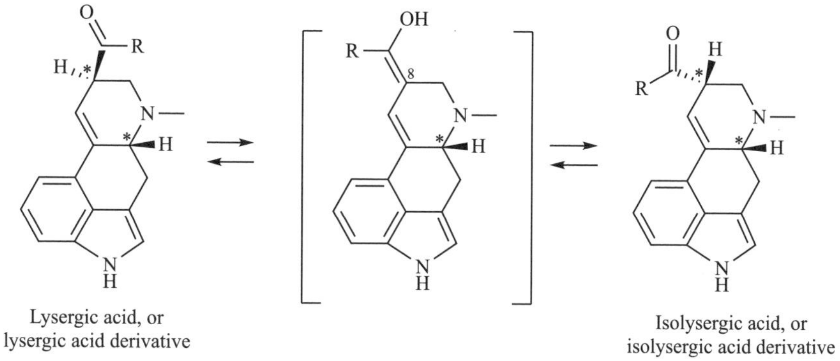
Lysergic acid, or
lysergic acid derivative
Isolysergic acid, or
isolysergic acid derivative

e) The stereochemistry of lysergic acids and dihydrolysergic acids

Through the investigations carried out in connection with the clarification of the lysergic acid–isolysergic acid isomerism, discussed in the previous section, the steric relationships between the various isomeric lysergic acids and dihydrolysergic acids had been clarified. In contrast, the question of the configurational relations between the centers of asymmetry within the molecules of the individual isomers was still open, i.e., it remained to be clarified whether the substituents at the two resp. three centers of asymmetry were *cis-* or *trans-* to each other. The final stereochemical problem that remained was the determination of the absolute configuration of lysergic acid.

To solve the first problem, conformational theory could be used[97, 133, 134]. The following derivations of configurational and conformational formulae of the isomeric dihydrolysergic acids and lysergic acids are based primarily on the detailed experimental investigations of *A. Stoll* and coworkers[97].

The evaluation of the following experimental findings according to the views of the conformational theory made it possible to assign the four isomeric dihydrolysergic acids the spatial formulae III, IV, V, and VI (Formula Scheme 2, p. 58), in which the absolute configuration, determined later, is anticipated.

a) Derivatives of lysergic acid yield only one reduction product during catalytic hydrogenation, which is always based on the dihydrolysergic acid-(I). On the other hand, two isomers are formed during the reduction of derivatives of isolysergic acid, which are derived from dihydroisolysergic acid-(I) and dihydroisolysergic acid-(II). In rapid hydrogenation with platinum catalyst, the iso-(II) compound predominates, while slower reduction with palladium gives rise to both isomers in approximately equal amounts.

b) Dihydroisolysergic acid-(I) methyl ester yields the epimeric dihydrolysergic acid-(I) upon vigorous alkaline hydrolysis.

c) Reactions at the carboxyl group proceed much faster for dihydrolysergic acid-(I) and dihydroisolysergic acid-(II) than for dihydroisolysergic acid-(I) and dihydrolysergic acid-(II). For example, dihydrolysergic acid-(I) azide reacts with amines faster than the iso-(I) compound. The same picture emerges from the hydrolyzation numbers of the four isomeric dihydrolysergic acid methyl esters listed in Table 5.

Table 5
Hydrolysis percentage of the isomeric dihydrolysergic acid methyl esters (6 hours at room temperature with excess 0.1 N alcoholic potassium hydroxide solution)

Dihydrolysergic acid-(I)	73.3%
Dihydroisolysergic acid-(I)	38.1%
Dihydrolysergic acid-(II)	14.3%
Dihydroisolysergic acid-(II)	52.6%

d) 6-Methyl-8-amino-ergoline-(I) yields the corresponding alcohol when treated with nitrous acid, while the epimeric 6-methyl-8-amino-isoergoline-(I) yields an unsaturated compound with elimination of the amino group.

e) On the aluminum oxide column, derivatives of dihydrolysergic acid-(I) migrate more slowly than those of dihydroisolysergic acid-(I) on one hand, and compounds of dihydroisolysergic acid-(II) migrate more slowly than dihydrolysergic acid-(II) derivatives on the other. 6-Methyl-8-acetoxy-isoergoline-(I) is easier to elute than 6-methyl-8-acetoxy-ergoline-(I).

The equatorial, as opposed to axial, arrangement of the substituent in the 8-position can be decided on the basis of these facts: dihydrolysergic acid-(I) and dihydroisolysergic acid-(II) carry the 8-substituent in equatorial arrangement (favored in epimerization reactions, relatively sterically unhindered in hydrolysis and condensation reactions, elimination of the 8-amino group difficult—upon exposure of the 8-amino group to HNO_2, it is replaced by a hydroxyl group—relatively high retention in the chromatogram). In dihydroisolysergic acid-(I) and dihydrolysergic acid-(II) the substituent has on the contrary an axial position (unstable in epimerization reactions, relatively sterically hindered, easy elimination of the 8-amino group on exposure to HNO_2, lower adhesion in chromatography).

Formula Scheme 2

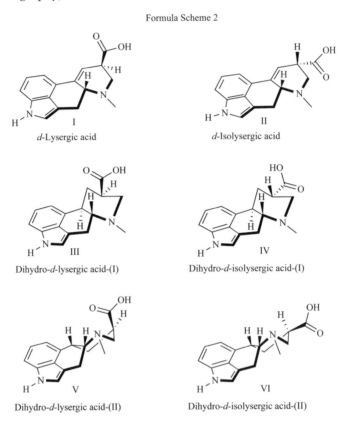

I
d-Lysergic acid

II
d-Isolysergic acid

III
Dihydro-d-lysergic acid-(I)

IV
Dihydro-d-isolysergic acid-(I)

V
Dihydro-d-lysergic acid-(II)

VI
Dihydro-d-isolysergic acid-(II)

To solve the question of the *cis*- or *trans*-linkage of rings C and D, fewer facts are available. To that end, the behavior of lysergic, resp. isolysergic acid during hydrogenation must be consulted. The predominant formation of dihydroisolysergic acid-(II) under conditions favoring the formation of *cis*-compounds (rapid hydrogenation with platinum) indicates *cis*-configuration for this isomer. The striking fact that during the hydrogenation of lysergic acid only one of the two conceivable dihydro derivatives, the dihydrolysergic acid-(I), is formed can be explained according to *R. C. Cookson*[133] that in this case the carboxyl group and the 5-hydrogen atom are located on the same side of the molecule and thus shielding the double bond on one side. The attack of the hydrogen during hydrogenation thus inevitably occurs from the other side of the molecule, resulting in the *trans*-compound. Thus, the linkage of rings C and D in two of the isomeric dihydrolysergic acids is defined on the basis of two mutually independent considerations, in both cases resulting in the same assignment.

The dihydrolysergic acid-(I) according to these deductions fulfills the C/D-*trans*-8-equatorial formula III, the epimeric dihydroisolysergic acid-(I) the C/D-*trans*-8-axial formula IV. The dihydrolysergic acid-(II) and dihydroisolysergic acid-(II) are due to their mode of formation and the above deductions assigned conformations V and VI.

Based on known correlations between IR spectra and conformational ratios in the steroid series, correlations between the IR spectra of different isomeric dihydrolysergic acid derivatives and their above-derived conformations could also be established[97].

The pK values of stereoisomeric ergoline derivatives, as compiled in Table 6, could also be related to the derived conformational formulae.

Table 6
pK values of stereoisomeric ergoline derivatives

	Acids		Methyl ester
	pK*$_1$	pK*$_2$	
Dihydrolysergic acid-(I)	4.85	7.85	6.20
Dihydroisolysergic acid-(I)	4.45	9.25	6.40
Dihydrolysergic acid-(II)	4.67	9.28	6.86
Dihydroisolysergic acid-(II)	4.97	8.38	6.42
6-methyl-8-hydroxy-ergoline-(I)	6.55		
6-methyl-8-hydroxy-isoergoline-(I)	7.80		

The compounds were dissolved in excess 0.1 N hydrochloric acid in 80% cellosolve. The isomeric dihydrolysergic acids and their methyl esters were back-titrated with 0.1 N sodium hydroxide, the isomeric 6-methyl-8-hydroxy-ergolines with 0.1 N tetramethylammonium hydroxide in the same solvent.

With regard to the relationship between pK values and the spatial distance between the carboxyl and amino groups in amino acids, contradictory views can be found in the literature. According to *J. B. Stenlake*[134], two factors, namely the inductive and the electrostatic effect, are generally decisive for the net basicity or acidity. However, since in the case of stereoisomeric amino acids the inductive effect propagated through the chain of carbon atoms remains the same in all cases, only the electrostatic interactions between the two groups caused by the spatial arrangement must be considered for comparison.

Due to spatial proximity, both the acidic character of the carboxyl group and the basicity of the amino group are enhanced. In compounds of the dihydrolysergic acid type, the condition of spatial proximity is fulfilled for the isomers with an axial carboxyl group; these must therefore be both stronger acids and stronger bases than the corresponding epimers with an equatorial carboxyl group. This is indeed the case, as can be seen from a comparison of epimeric pairs in Table 6. In accordance with the rule formulated above, the two isomers with axial carboxyl group, dihydroisolysergic acid-(I) and dihydrolysergic acid-(II), are both stronger acids and stronger bases than the corresponding epimers with equatorial substituent. The negative polarization of the carbonyl oxygen also has the same effect on the methyl esters. In agreement with findings with amino alcohols of the tropane series[135, 136], the axial 6-methyl-8-hydroxy-isoergoline-(I) proves to be a significantly stronger base than the epimer with an equatorial OH group.

Based on the steric ratios derived above for the dihydrolysergic acids, the configurations I and II ensue for lysergic acid and isolysergic acid, i.e., in the lysergic acid the hydrogen atoms at C5 and C8 are in the *trans* position, in the isolysergic acid they are in the *cis* position. Within these configurations, however, different conformations are still possible, depending on whether the ring D has a pseudo boat shape or a pseudo chair shape. If the pseudo chair shape is present, as shown in Formulae I and II, then the carboxyl group in the lysergic acid is in an equatorial position. In the pseudo chair form, the carboxyl group in the lysergic acid would come to lie in an axial position. In order to decide which of these two forms the ring D actually has, it was necessary to clarify which position, equatorial or axial, the carboxyl group in the lysergic acid and isolysergic acid takes. This could be decided mainly by determining the

isomerization equilibria between lysergic acid and isolysergic acid derivatives (Tables 7 and Figure 9) and comparing the pK values of stereoisomeric lysergic acid amides (Table 8).

Table 7
Isomerization equilibria between lysergic acid and
isolysergic acid derivatives

		Lysergic acid form %	Isolysergic acid form %
· · · · · · · · ·	Ergosine	42	58
· — · — ·	Ergocryptine	48	52
———————	Ergobasine	52	48
· · — · · —	Lysergic acid ethylamide	54	46
Δ – Δ – Δ	Lysergic acid dimethylamide	84	16
– – – – –	Lysergic acid diethylamide	88	12

From the curves shown in Figure 9 and the equilibrium numbers
listed in Table 7, it is evident that there is a significant difference between
lysergic acid derivatives of primary amines on the one hand and those
of secondary amines on the other, in that for the former the equilibria
are generally around 50:50%, while for the latter the lysergic acid form
strongly predominates.

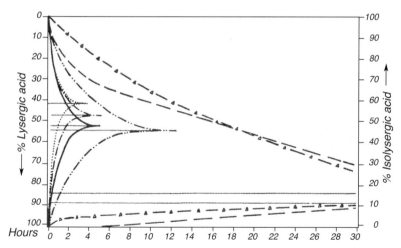

Figure 9. Time course of the rearrangement of lysergic acid and isolysergic acid
derivatives in 0.1 N methanolic potassium hydroxide solution
Left: Percentage content of the isomer mixture of lysergic acid form
Right: Percentage content of the isomer mixture in the isolysergic acid form

Table 8

pK values of stereoisomeric lysergic acid amides

	pK	Δ
Ergobasine	$6.30^{1)}$; $6.79^{2)}$	$0.25^{1)}$; $0.64^{2)}$
Ergobasinine	$6.55^{1)}$; $7.43^{2)}$	
Lysergic acid ethylamide	$6.09^{1)}$	0.26
Isolysergic acid ethylamide	$6.35^{1)}$	
Lysergic acid diethylamide	$6.37^{1)}$	1.15
Isolysergic acid diethylamide	$7.52^{1)}$	
Lysergic acid dimethylamide	$6.39^{1)}$	1.03
Isolysergic acid dimethylamide	$7.42^{1)}$	

[1] In 80% ethanol.
[2] In water.

The pK values listed in Table 8 show that, similar to the isomerization equilibria, there are characteristic differences between the monoalkylamides on the one hand and the dialkylamides on the other. While the differences between the pK values of the normal and the isoform of the monoalkylamides are relatively small, the dialkylamides of the isolysergic acid prove to be significantly stronger bases than their epimers.

In view of the large differences in epimerization and pK values between monoalkyl and dialkylamides of lysergic acid and isolysergic acid, the question arises, first of all, as to which of the data given in Tables 7 and 8 are significant for the assessment of the conformation. These differences can be attributed to the presence or absence of a hydrogen atom on the amide nitrogen in the two types of compounds. In the case of monoalkylamides, the formation of a hydrogen bond between the carbonyl amido group and the basic nitrogen atom must be expected if the steric conditions are favorable for this.

One may therefore well consider the proportions of the dialkylamides, in which this complication cannot occur, as the normal ones.

The position of the equilibrium in the epimerization of the lysergic and the isolysergic acid diethylamide or dimethylamide suggests greater thermodynamic stability of the lysergic acid form. From the pK values of the two pairs of compounds, a comparatively smaller distance between amidocarbonyl group and basic nitrogen in the iso compounds can be concluded, based on the considerations employed for the dihydro derivatives. Both findings can be interpreted by presuming an equatorial position of the amidocarbonyl group in the lysergic acid derivative, resp. an axial arrangement in the isolysergic acid derivative. This is the case in the pseudo armchair forms, and thus Formula I corresponds to lysergic acid and Formula II to isolysergic acid.

The above discussed derivations refer only to the relative spatial position at the various centers of asymmetry of the isomeric lysergic acids and dihydrolysergic acids, i.e., it has not yet been clarified whether the spatial formulae recorded in Formula Scheme 2 reflect the spatial relationships of the natural lysergic acid, designated as *d*-form, and its derivatives, or its antipode, the only artificially accessible *l*-lysergic acid. Determination of the absolute configuration of lysergic acid[98] made it possible to describe these formulae to correspond to *d*-lysergic acid.

The rotatory dispersion method was used to determine the absolute configuration[137].

A comparison of the rotatory dispersion curves of the four isomeric lysergic acids (Figure 10) shows that solely the steric ratios at carbon atom 5 (linkage of rings C and D) are decisive for a positive or negative Cotton effect and that the configuration at the second center of asymmetry, at carbon atom 8, has merely an additive or subtractive effect. This finding is in agreement with the observations made by *C. Djerassi*[137] on steroids, according to which centers of asymmetry involved in ring linkages exert the decisive influence on the spectropolarimetric curve.

It was now possible to configurationally relate model substances containing the rings C and D, resp. A, C, and D, of lysergic acid, with polycyclic compounds from the steroid series, whose absolute configuration was known, with respect to the spatial arrangement at C5 of the lysergic acid.

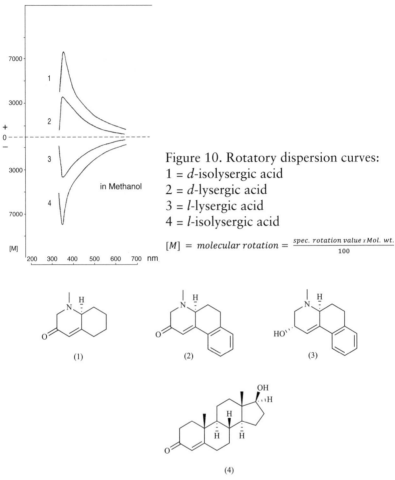

Figure 10. Rotatory dispersion curves:
1 = *d*-isolysergic acid
2 = *d*-lysergic acid
3 = *l*-lysergic acid
4 = *l*-isolysergic acid

$$[M] = molecular\ rotation = \frac{spec.\ rotation\ value \times Mol.\ wt.}{100}$$

(1)   (2)   (3)

(4)

$(+)\text{-}\Delta^{4,4a}$-N-methyl-octahydroquinolin-3-one (formula 1), $(+)\text{-}\Delta^{1,10b}$-N-methyl-hexahydrobenzo[f]quinolin-2-one (formula 2), and the corresponding quinolin-2-ol (formula 3), like testosterone (formula 4), show a positive Cotton effect (cf. Figures 11-13), which justifies assigning to these compounds the same configuration at the center of asymmetry corresponding to C5 of lysergic acid as in testosterone, namely with a ß-position hydrogen atom. Since the tricyclic alcohol base (3), apart from the missing pyrrole ring and the functional group at C8, which, however, have no influence on the direction of rotation of the rotational dispersion, corresponds in structure to lysergic acid, it is permissible to conclude

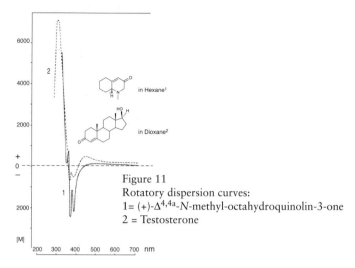

Figure 11
Rotatory dispersion curves:
1= (+)-$\Delta^{4,4a}$-N-methyl-octahydroquinolin-3-one
2 = Testosterone

from the analogous course of the rotational dispersion curves of (3) on the one hand (cf. Figure 13) and of d-lysergic acid and d-isolysergic acid on the other hand (cf. Figure 10) an accordant configuration. This means that in natural lysergic acid the hydrogen atom at C5 is in the ß-position, as expressed in the configuration formulae of the following Formula Scheme 3 and in the spatial formulae of Scheme 2 (p. 58), which thus represent the absolute configurations of the isomeric lysergic acids and dihydrolysergic acids presented there.

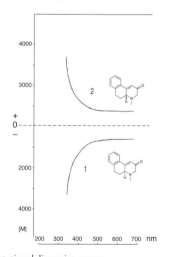

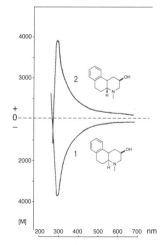

Fig. 12. Rotational dispersion curves:
1 = (-)-$\Delta^{1,10b}$-N-methyl-hexahydrobenzo[f]quinolin-2-one
2 = (+)-$\Delta^{1,10b}$-N-methyl-hexahydrobenzo[f]quinolin-2-one

Fig. 13. Rotational dispersion curves:
1 = (-)-$\Delta^{1,10b}$-N-methyl-hexahydrobenzo[f]quinolin-2-ol
2 = (+)-$\Delta^{1,10b}$-N-methyl-hexahydrobenzo[f]quinolin-2-ol

Formula scheme 3

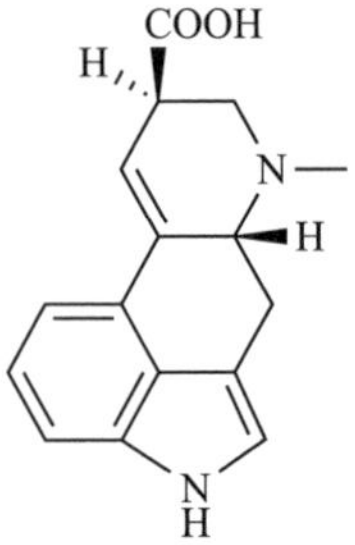

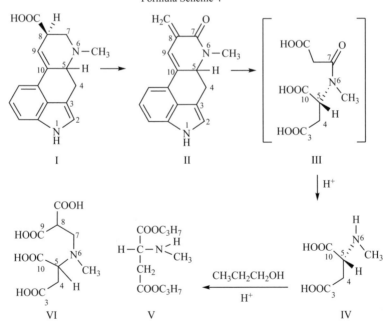

d-Lysergic acid 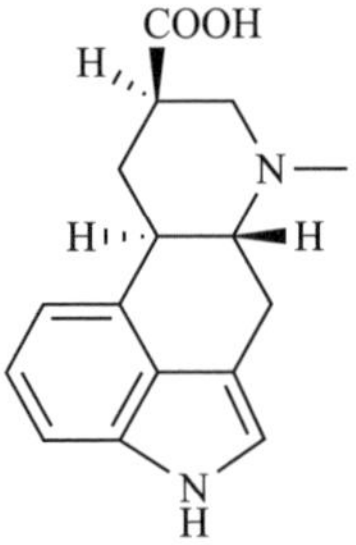   Dihydro-d-lysergic acid-(I) 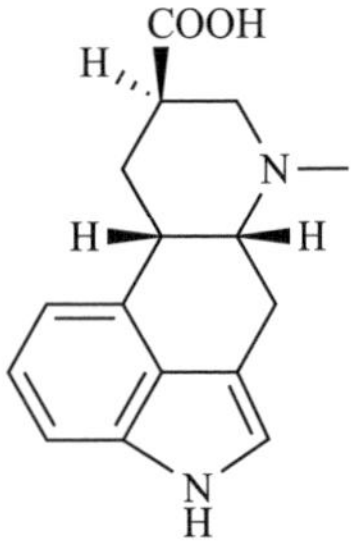   Dihydro-d-lysergic acid-(II)

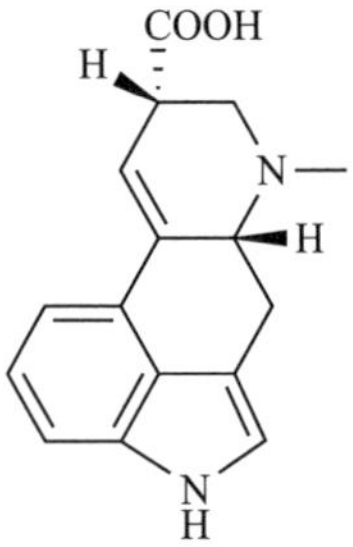

d-Isolysergic acid 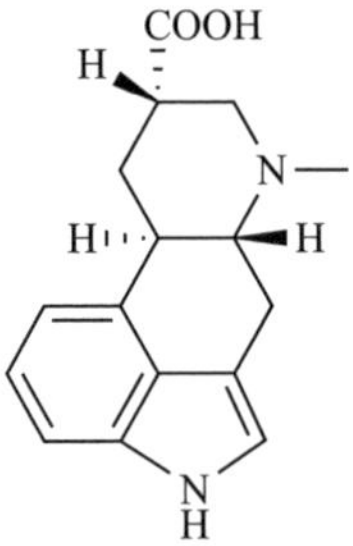   Dihydro-d-isolysergic acid-(I) 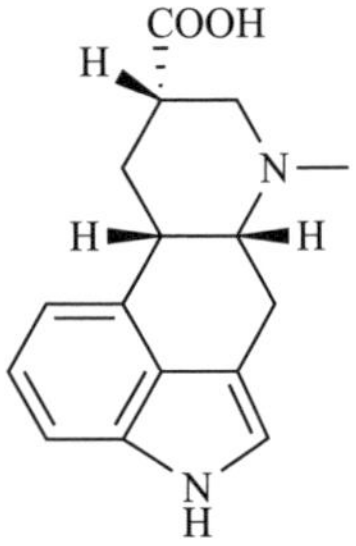   Dihydro-d-isolysergic acid-(II)

Formula Scheme 4

I   II   III

VI   V   IV

The absolute configuration of lysergic acid derived on the basis of the rotatory dispersion curves was confirmed chemically, by oxidatively degrading d-lysergic acid lactam to a derivative of D-aspartic acid by the route shown in Formula Scheme 4 above[99]. The d-lysergic acid could not be used as such for the degradation because it would give rise to a nitrogen-dialkylated aspartic acid (VI), which was known to racemize rapidly. To obtain a sterically stable degradation product with a secondary nitrogen atom, N-nor-lysergic acid would have to be used as the starting material. However, since the demethylation of lysergic acid was not successful, lactam II was used for the degradation, which is formed when lysergic acid is briefly heated in acetic anhydride[92]. It was ozonized in methylene chloride-methanol-water solution and then oxidized with hydrogen peroxide with the addition of anhydrous formic acid. Attempts to isolate the expected tricarboxylic acid III were unsuccessful, so the reaction mixture was treated with 1 N HCl to split off the oxalyl residue. After subsequent esterification of IV with n-propanol, an optically active amino acid ester was obtained in a yield of 4% in pure form. It proved to be identical with an authentic D-(+)-N-methylaspartic acid di-n-propyl ester (V). Natural lysergic acid thus has the absolute configuration expressed by Formula I, which had already been deduced from comparative rotatory dispersion studies. According to the nomenclature introduced by *R. S. Cahn, C. K. Ingold*, and *V. Prelog* to designate the absolute configuration[138], the natural d-lysergic acid would thus be designated as (5R,8R)-lysergic acid.

f) Syntheses of dihydrolysergic acids

The path to the total synthesis of lysergic acid involved several steps, as was to be expected with such a complicated, novel type of compound. The first fundamental step was the synthesis of the saturated, tetracyclic ring system, ergoline, from which all ergot alkaloids are derived. This was followed by the synthesis of racemic dihydrolysergic acid (= 6-methyl-8-carboxy-ergoline), then of the optically active dihydrolysergic acids, and finally of the natural lysergic acid (6-methyl-8-carboxy-$\Delta^{9,10}$-ergolene).

Prior to the syntheses of the dihydrolysergic acids, the historically important synthesis of ergoline by *W. A. Jacobs*[101] will therefore be briefly discussed.

Synthesis of ergoline

As shown in the Formula Scheme 5 below, the rings A and C of the ergoline system started in the form of 3-nitro-α-naphthoic acid (I). After reduction of the nitro group, ring D was added according to the method of *Skraup*'s quinoline synthesis. Nitration, reduction of the nitro group and lactam formation led to the naphthostyril derivative (V), in which all four rings of ergoline are already present. Partial hydrogenation of the pyridine ring and subsequent reduction with sodium in butanol yielded besides the amino alcohol (VIII) also the indole derivative (VII), the ergoline.

The starting material for the first synthesis of dihydrolysergic acid by *F. C. Uhle* and *W. A. Jacobs*[91] (see Formula Scheme 6) was 4-aminonaphthostyril (I), in which the rings A, B, and C of the ergoline system are already preformed. Condensation with cyanomalondialdehyde followed by cyclization with zinc chloride and hydrolysis with hydrochloric acid gave the tetracyclic ring skeleton of lysergic acid substituted in the 8-position with the carboxyl group (II). Catalytic hydrogenation of the chloromethylate of II, which was obtained via the iodomethylate, led to the tetrahydro-N-methyl compound (III). Reduction of this naphthostyril derivative with sodium in butanol yielded racemic dihydrolysergic acid (IV), which proved to be identical to a preparation obtained by hydrogenation of racemic lysergic acid of natural origin.

Formula Scheme 5

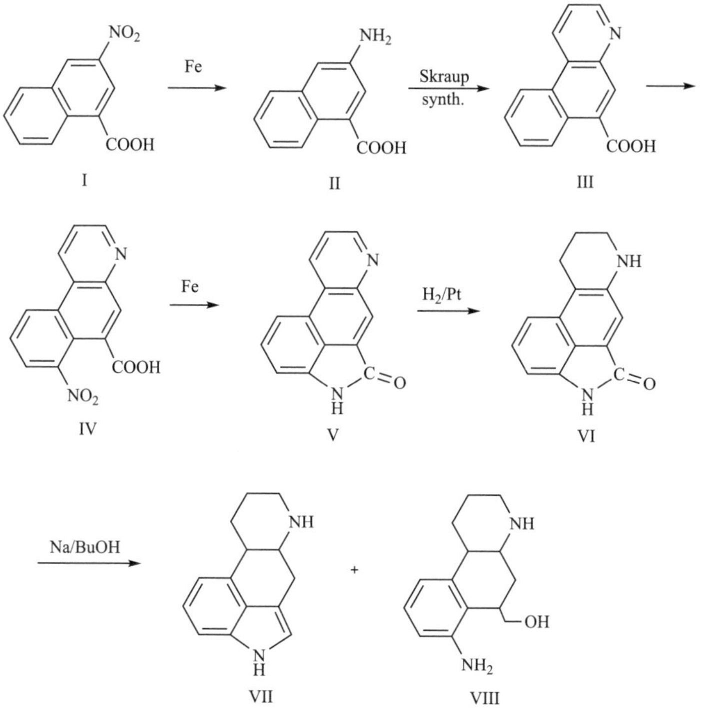

Formula Scheme 6

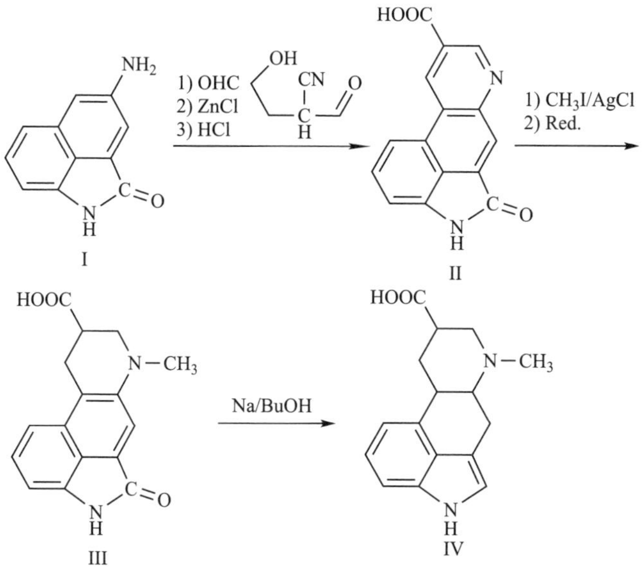

The naphthostyril derivative (III) has also been prepared later by other methods[139, 140], which opens up further synthetic routes to dihydrolysergic acid. Furthermore, the tetracyclic naphthostyril derivative (II) has been constructed by an alternative route, starting from 1-hydroxymethylene-1-phenyl-2-propanone as ring A, by successive addition of rings D, C, and B[141].

A new improved synthesis of dihydrolysergic acid, which leads to the uniform racemates as well as to the optically active dihydrolysergic acids, proceeds via the racemic dihydro-nor-lysergic acids.

Synthesis of the four isomeric racemic dihydro-nor-lysergic acids[142]

A. Stoll and *J. Rutschmann* used as a starting material for the synthesis of the dihydro-nor-lysergic acids (see Formula Scheme 7) a compound which had already been described earlier by *R. G. Gould Jr.* and *W. A. Jacobs*[143], namely the 3'-amino-3-carbethoxy-5,6-benzoquinolin-4-one-7-carboxylic acid lactam (IV). This substance is prepared by

Formula Scheme 7

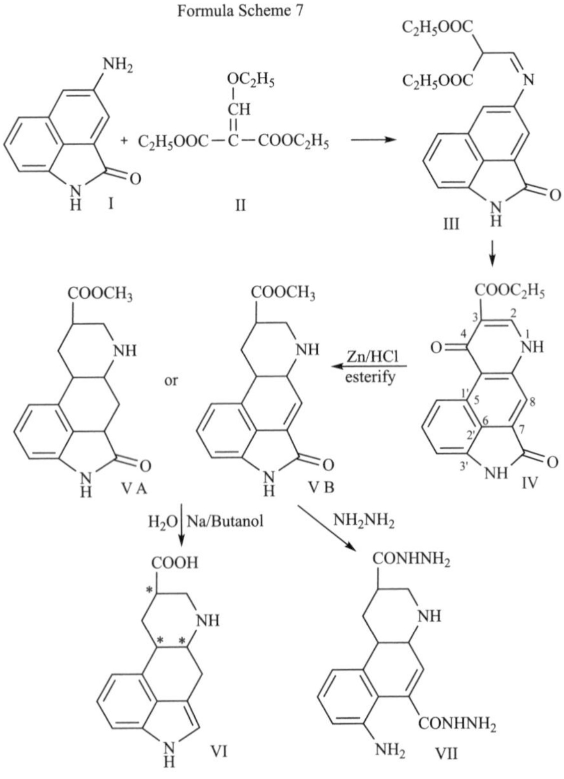

70

condensation of 3-aminonaphthostyril (I) with diethyl 2-(ethoxymeth-ylene)malonate (II) and a ring closure of the thus obtained N-(3-naph-tostyril)-ß-amino-α-carbethoxyacrylic ethyl ester (III). Compound (IV) is conveniently accessible in larger quantities and already contains the complete skeleton of lysergic acid. While *R. G. Gould* and *W. A. Jacobs*, by the reduction of (IV) with amalgamated zinc in glacial acetic acid, were able to achieve only a hydrogenation in the 2–3 position, it was possible to reduce ring D to the piperidine stage (V) by reduction according to *Clemmensen* in boiling glacial acetic acid with amalgamated zinc and hydrochloric acid. The acidic reduction product was converted to the methyl ester for isolation. Since the compound did not crystallize, it could not be decided to what extent ring C had been reduced; it is likely to have either the V A or V B structure.

This formulation is in agreement with the fact that substance V A or B, in contrast to the yellow γ-quinolone IV, is colorless, i.e., no longer possesses the chromophore system of the naphthostyril. Furthermore, by a treatment with hydrazine, a crystallized dihydrazide could be produced which, according to the analysis, should have the structure VII.

By treating V with sodium in butanol, the molecules could be fur-ther reduced. When adhering to certain conditions, at least in part, the desired conversion to the indole derivative VI, i.e., to dihydro-nor-lyser-gic acid, occurred. For this reaction to succeed, it is important that the butanol contains a trace amount of water, so that the methyl ester group is hydrolyzed before its undesired reduction to the alcohol, according to *Bouveault-Blanc*, occurs.

The reaction mixture was worked up by esterifying the products soluble in aqueous alkali with methanol after the removal of the inor-ganic sodium compounds. The ester fraction was chromatographed by a gradient flow method on aluminum oxide using benzene as solvent to which increasing amounts of chloroform were added. Three crystal-lized fractions were obtained, all of which had the same gross formula

$C_{16}H_{18}O_2N_2$, and their further examination showed that the methyl esters of three isomeric racemic dihydro-nor-lysergic acids were present.

By hydrolyzing the esters with aqueous alcoholic lye, the three racemic dihydro-nor-lysergic acids were obtained. These were well-crystallizing, amphoteric compounds which largely resemble the dihydrolysergic acids in their solubility properties. Reaction of the methyl esters with hydrazine produces hydrazides, and treatment of the esters with acetic anhydride produces the particularly characteristic N-acetyl derivatives, which can be hydrolyzed to N-acetyl dihydro-nor-lysergic acids when boiled with aqueous alcoholic lye.

All these compounds give a violet-blue *Keller*'s color reaction indistinguishable from those of natural dihydrolysergic acids.

It was natural to assume that the three isomeric dihydro-nor-lysergic acids would sterically correspond to the three stereoisomeric natural dihydrolysergic acids, since the relative ratios in both groups are the same.

The proof that one nor-acid belongs to the dihydrolysergic acid series arose from the identity of the reaction product of its methyl ester with methyl iodide and the racemic dihydrolysergic acid-(I) methyl ester iodomethylate. This acid is therefore known as racemic dihydro-nor-lysergic acid-(I).

In another way, a second nor-acid could be related to one of the three natural dihydrolysergic acids. The alkaline hydrolysis of the hydrazide of this dihydro-nor-acid did not produce the same nor-acid as the hydrolysis of its methyl ester, but dihydro-nor-lysergic acid. Thus, a rearrangement in the dihydrolysergic acid series has occurred. This rearrangement is characteristic of the dihydroisolysergic acid-(I) structure and has so far only been observed with it. This nor-acid is therefore designated racemic dihydro-nor-isolysergic acid-(I).

It was obvious to assign the remaining nor-acid to the third known dihydrolysergic acid, dihydroisolysergic acid-(II), and to call it racemic dihydro-nor-isolysergic acid-(II).

The relative retention of the esters of the nor- and the natural dihydrolysergic acids on the aluminum oxide column is also in agreement with this classification. If the dihydro-nor-lysergic acid methyl esters are numbered according to decreasing retentive strength, the assignment just made gives the series dihydro-nor-isolysergic acid-(I) methyl ester[121], dihydro-nor-lysergic acid methyl ester, dihydro-nor-isolysergic acid-(II) methyl ester, which parallels the adsorptive behavior of the corresponding esters methylated on the nitrogen, which are eluted in the order dihydroisolysergic acid-(I) methyl ester, dihydrolysergic acid methyl ester, dihydroisolysergic acid-(II) methyl ester.

By alkaline hydrolysis of the racemic dihydro-nor-isolysergic acid-(II) hydrazide, it was possible to achieve partial epimerization at C8 and in this way to prepare the fourth, theoretically possible isomeric racemate, the racemic dihydro-nor-lysergic acid-(II)[144]. While in the corresponding rearrangement of dihydroisolysergic acid-(I) into dihydrolysergic acid-(I)[121] the equilibrium is entirely on the side of the latter isomer, in the hydrolysis of the racemic dihydro-nor-isolysergic acid-(II) hydrazide only about 20% dihydro-nor-lysergic acid-(II) hydrazide is obtained alongside about 80% dihydro-nor-isolysergic acid-(II).

Some characteristic data of the four isomeric racemic dihydro-nor-lysergic acids, their methyl esters, 6-acetyl methyl esters and hydrazides are summarized in Table 9.

Table 9
Properties of the four stereoisomeric dihydro-nor-lysergic acids
and their simple derivatives

	DH-nor-LA-(I) racemic	DH-nor-LA-(II) racemic	DH-nor-iso-LA-(I) racemic	DH-nor-iso-LA-(II) racemic
Free acids	m.p. over 350° polyg. leafletts from water	m.p. 237-240° fine needles from water	m.p. over 350° thin prisms from water	m.p. 320-330° flat needles from water
Methyl ester	m.p. 204-206° prisms from acetic acid	m.p. 151-152° fine needles from benzene-chloroform	m.p. 161-162° rhombic polyhedra from benzene	m.p. 75-80° needles from methanol-water
6-Acetyl-methyl ester	m.p. 280-282° mass. polyhedra from methanol	m.p. 247-250° rhombic plates from benzene	m.p. 205-206° mass. polyhedra from benzene	m.p. 167-168° fine cryst. from methanol-water
Hydrazide	m.p. 280-281° small druses from methanol-water	m.p. 288-289° prisms from methanol	m.p. 241-242° leaflets from methanol	m.p. 139-140° needles from methanol-water

Further variants of dihydro-nor-lysergic acid synthesis

The syntheses of the racemic dihydro-nor-lysergic acids and racemic dihydrolysergic acid have the common feature that the ring D of the dihydrolysergic acid is attached to the naphthostyril ring skeleton, which is then reduced to the indole system in the final step by vigorous means[145], [146]. The two following variants (see Formula Scheme 8) differ from this in that here a naphthostyril derivative (I) is first hydrogenated to the corresponding benz[c,d]indoline (II), after which the assembly of ring D was carried out using the methods described earlier.

Formula Scheme 8

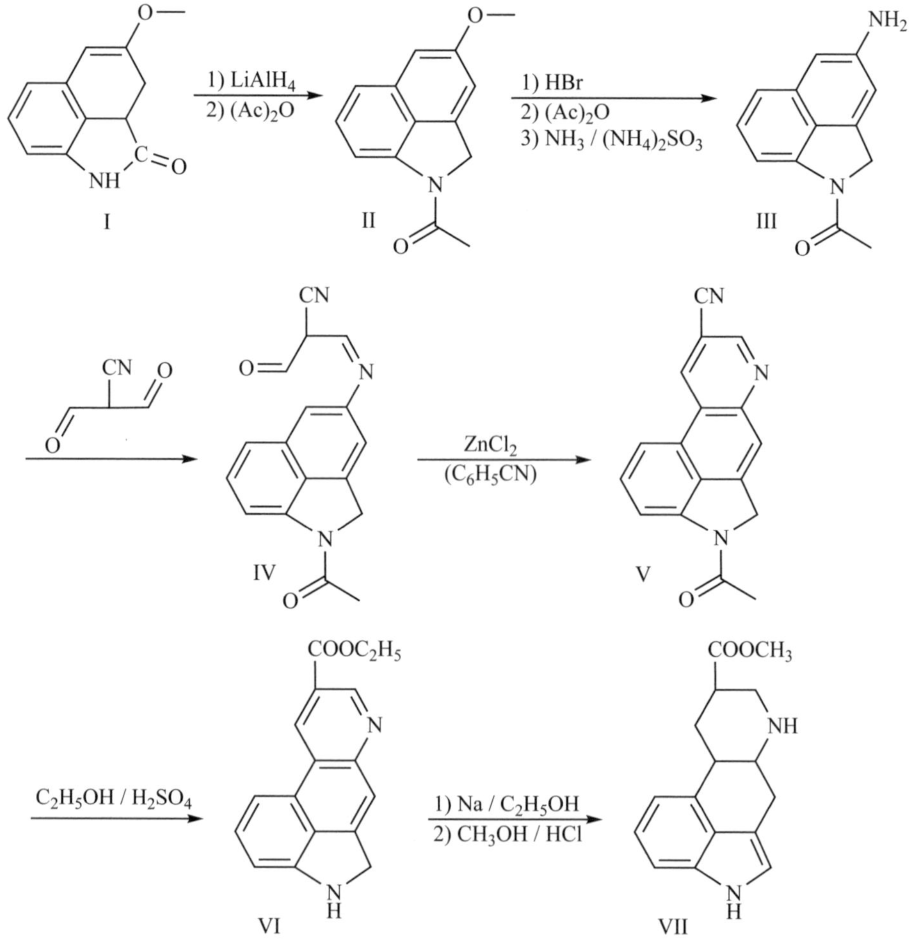

According to one variant (see Formula Scheme 8), 1-acetyl-4-aminobenz[c,d]indoline (III) was condensed with cyanomalonicdialdehyde according to the instruction by *F. C. Uhle* and *W. A. Jacobs*[91] and converted via stages IV–VI into the benzindoline derivative VI. Reduction of VI with sodium in ethanol provided in a low yield the corresponding indole compound, a mixture of the isomeric racemic dihydro-nor-lysergic acids, which could be characterized in the form of the methyl esters VII.

Formula Scheme 9

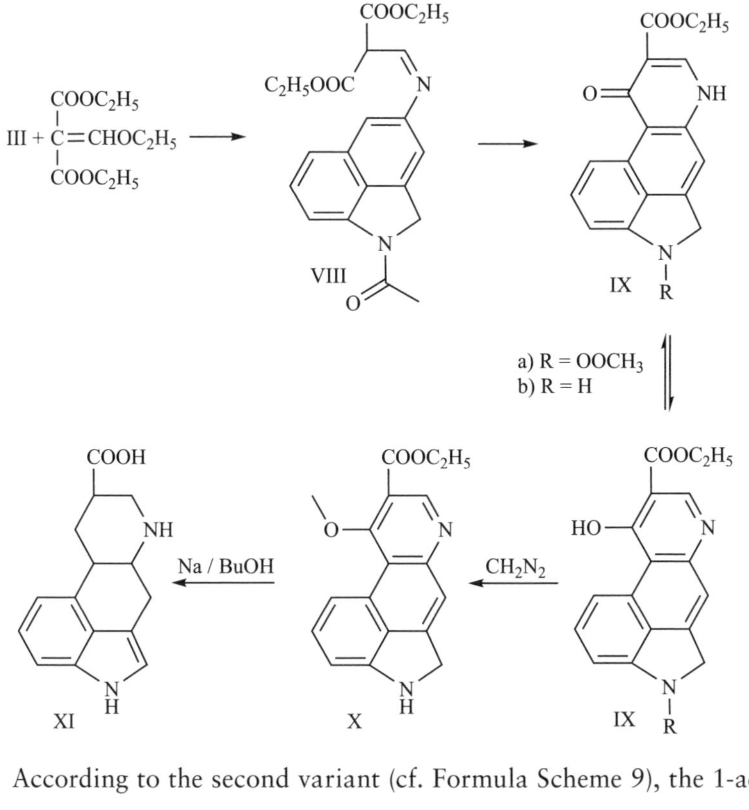

a) R = OOCH$_3$
b) R = H

According to the second variant (cf. Formula Scheme 9), the 1-acetyl-4-aminobenzindoline III was reacted with diethyl 2-(ethoxymethylene)malonate. 1-Acetyl-4-(2',2'-diethoxycarbonyl-ethylideneamino)benz[c,d]indoline (VIII) was formed in good yield and cyclized to the indolino-quinolone (IX) on heating in biphenyl. The hope of getting from this compound by a suitable reduction a dihydro-nor-lysergic acid with a hydroxyl functional group in the 9-position, from which the lysergic acid system could be formed by water cleavage, was not fulfilled. In the assumption that the N-acetyl group of compound IX could make the subsequent conversion into the lysergic acid structure more difficult, it was selectively removed. Since reduction attempts with the strongly yellow colored, poorly soluble compound IX b were unsuccessful, the 9-hydroxyl group was methylated with diazomethane, yielding the readily soluble substance X. Reduction with sodium and butanol gave, though in very low yield, indole compounds that would be identified as dihydro-nor-lysergic acids. The methoxyl group in the 9-position has thus been reductively eliminated.

Synthesis of the four isomeric racemic dihydrolysergic acids

Starting from the racemic dihydro-nor-lysergic acids described in the previous section, A. Stoll and coworkers were able to prepare the corresponding isomeric racemic dihydro-lysergic acids by methylation of the secondary nitrogen atom in ring D[93, 144, 147].

The methylation could be carried out by two different methods. Firstly, by transmethylation of the dihydro-nor-lysergic acid methyl esters, which, when heated above the melting point to about 220° in an evacuated vessel, are converted into the corresponding free dihydrolysergic

Table 10
Properties of isomeric racemic dihydrolysergic acids,
their methyl esters and hydrazides

	d,l-DH-LA-(I)	d,l-DH-LA-(II)	d,l-DH-iso-LA-(I)	d,l-DH-iso-LA-(II)
Acids	approx. 350° polyg. leafletts from water	273-275° small druses from methanol-water	approx. 317° polyg. flakes from water	approx. 315° rhombic sheet from water
Methyl ester	148-150° leafletts from benzene	168-171° prisms from benzene	156-158° prisms from methanol	amorphous
Hydrazide	259° polyg. plates from methanol	243-245° needles from methanol	296° elongated leaflets from methanol-water	227-229 ° needles from methanol

acids by migration of the methyl group from the carboxyl to the nitrogen. Alternatively, the nitrogen in position 6 of the dihydro-nor-lysergic acid esters could be methylated using formaldehyde and catalytic reduction. According to this method, larger alkyl groups can also be introduced by using higher aldehydes. The indole system remains unaffected when working with anhydrous aldehydes in neutral solution. The reaction can be carried out smoothly at normal pressure and room temperature with Raney nickel as catalyst.

Table 10 lists the properties of the four isomeric racemic dihydrolysergic acids and their methyl esters and hydrazides.

The homologues of racemic dihydrolysergic acid, obtained by reductive condensation of racemic dihydro-nor-lysergic acid methyl esters with higher aldehydes, as well as their methyl esters are listed in Table 11.

Table 11

Properties of racemic dihydro-lysergic acids and their homologues, and the methyl esters of these compounds

	6-alkyl-dihydro-nor lysergic acids, rac.	6-alkyl-dihydro-nor-lysergic acid methyl esters, rac.	
R = methyl	m.p. ca. 310° polyg. leaflets from water	m.p. 148-150° leaflets from benzene	pK_B 7.80
R = ethyl	m.p. 305° rhomb. leaflets from water	m.p. 153-154° leaflets from benzene-cyclohexane	pK_B 7.55
R = n-propyl	m.p. 279-280° rhomb. leaflets from water	m.p. 184-186° leaflets from ethyl acetate	pK_B 7.75
R = n-butyl	m.p. 253° quadr. leaflets from methanol	m.p. 175-156° quadr. plates from methanol	pK_B 7.88
R = n-pentyl	m.p. 261-263° quadr. leaflets from methanol	m.p. 151-153° quadr. leaflets from ethyl acetate	pK_B 7.83
R = n-hexyl	m.p. 251° quadr. leaflets from methanol	m.p. 107-108° no crystals from cyclohexane	pK_B 7.85

Synthesis of the optically active dihydrolysergic acids

By resolving the racemic dihydrolysergic acid prepared by total synthesis into the optical antipodes, *A. Stoll, J. Rutschmann,* and *W. Schlientz*[93] obtained the optically active dihydrolysergic acids.

Synthetic racemic dihydrolysergic acid azide, which had been prepared in the usual manner from the ester via the hydrazide, was acid-amide linked with L-norephedrine. The resulting partial racemate could be separated into the two components on an aluminum oxide column using ethyl acetate and acetone as eluent. The less retained isomer showed a specific rotation of $[\alpha]_D^{20} = -114°$ (in pyridine) and crystallized from acetone in massive prisms and polyhedra of m.p. 240–241°. It was identical to the acid amide prepared from dihydro-*d*-lysergic acid azide of natural origin and L-norephedrine. The more firmly adhering isomer, the L-norephedrine of dihydro-*l*-lysergic acid, had a spec. rotation value of and crystallized from ethyl acetate in fine needles which melted at 252–253°.

Vigorous alkaline hydrolysis of the norephedrides provided the optically active dihydrolysergic acids. The antipode with the specific rotation value and a m.p. above 300° was identical with (-)-dihydro-*d*-lysergic acid, which is the basis of the dihydrogenated natural levorotatory ergot alkaloids. Its antipode with the spec. rotation value $[\alpha]_D^{20} = +107°$ (in pyridine) is to be understood as (+)-dihydro-*l*-lysergic acid. The optically active synthetic preparations crystallized from water like the natural dihydrolysergic acids in the characteristic rhombic or hexagonal platelets.

g) Synthesis of lysergic acid

Various research groups have been engaged in studies on the synthesis of lysergic acid, which have not led to success. These works can therefore not be reviewed here. The reference to the original literature must suffice[148–156, 370–376].

The first and to date still the only synthesis of lysergic acid was carried out by the American research group of *E. C. Kornfeld, E. J. Fornefeld, G. B. Kline, M. J. Mann, R. G. Jones, D. E. Morrison,* and

R. B. Woodward[95, 96]. Whereas the above-described routes for the formation of dihydrolysergic acid lead via the naphthostyril or via the benz[c,d] indoline system, from which the incorporation of the double bond characteristic of lysergic acid in the 9–10 position was not successful, this synthesis starts from a 2,3-dihydroindole derivative (see Formula Scheme 10). This prevented the formation of the benzindoline system (cf. Section C, III/2h) and allowed the application of classical methods for the construction of the unsaturated ring D of lysergic acid. Only in the last step was the 2,3-dihydroindole derivative dehydrogenated to the indole compound, lysergic acid.

Formula scheme 10

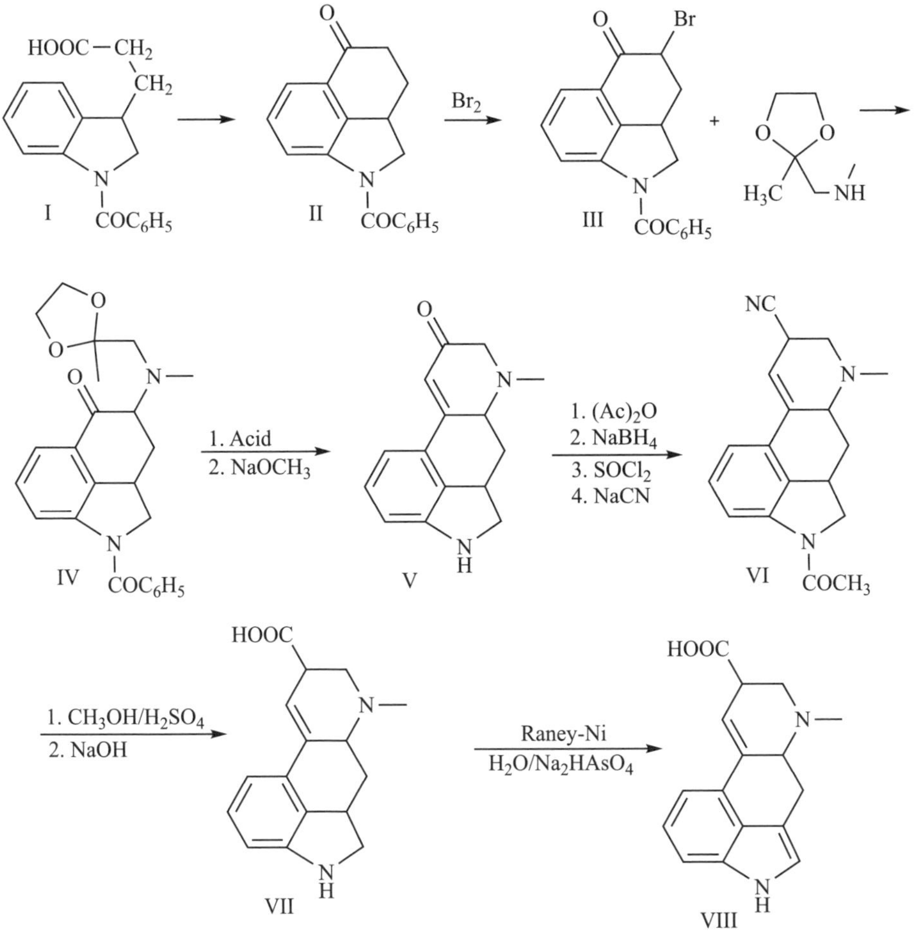

79

N-Benzoyl-3-indolepropionic acid (I) was condensed via the acid chloride to the tricyclic ketone (II). The bromination to the α-bromo ketone (III) was followed by the reaction with N-methyl-1-(2-methyl-1,3-dioxolan-2-yl)methanamine to compound IV. After acid hydrolysis to the methyl ketone, this could be reacted with sodium methylate to the tetracyclic unsaturated ketone, 9-keto-7-methyl-4,5,5a,6,6a,7,8,9-octahydroindolo[4,3-fg]quinoline (V). The released secondary amino group was again protected by acylation and hereupon the ketone was converted into the secondary alcohol by reduction with sodium borohydride and the latter into the corresponding chloride in sulfur dioxide solution with thionyl chloride. Reaction with sodium cyanide in liquid hydrocyanic acid yielded the nitrile (VI), which was converted by methanolysis into the ester, which was alkali hydrolyzed to 2,3-dihydrolysergic acid (VII). Dehydrogenation with deactivated Raney nickel in aqueous solution in the presence of sodium arsenate yielded racemic lysergic acid (VIII). This was separated into the optical antipodes according to a previously described procedure[117] in the form of the isolysergic acid hydrazide, thereby recovering the d-lysergic acid underlying the natural ergot alkaloids. Since the conversion of the racemic isolysergic acid hydrazide into ergobasine (ergometrine, ergonovine) had already been carried out earlier by A. Stoll and A. Hofmann[157], not only the synthesis of the lysergic acid but also, for the first time, the total synthesis of a natural ergot alkaloid has been realized. This synthesis has not yet found industrial application, since various steps are difficult to carry out on a technical scale and, above all, since the dehydrogenation from the indoline compound into the indole compound proceeds only with a low yield.

h) Rearrangement of the lysergic acid into the benz[c,d]indoline isomer

As discussed in Section III/2d, lysergic acid and its derivatives isomerize in hydroxyl-containing solvents to isolysergic acid or isolysergic acid compounds. In contrast to this reversible rearrangement, lysergic acid, isolysergic acid, and their derivatives undergo under the influence of strongly acidic reagents an irreversible isomerization, which consists of the transition of the indole system into the benz[c,d]indoline structure.

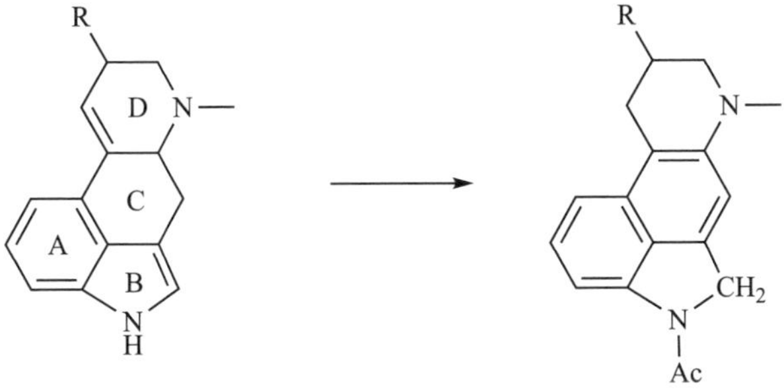

Ia R = COOH or COOCH₃
Ib R = CH₂OAc

IIa R = COOH, COOCH₃ or COOC₂H₅
IIb R = CH₂OAc

Hydrochloric acid in glacial acetic acid, chloroform, methanol, or alcohol has proven to be a suitable proton donor for the rearrangement of lysergic acid or lysergic acid esters. The course of the isomerization can be followed by the disappearance of the color reaction according to *van Urk-Smith,* since the indoline derivatives formed with this reagent no longer give any color.

In agreement with findings observed on benz[c,d]indolines obtained by LiAlH₄ reduction of naphthostyril derivatives[158], the acid rearrangement products proved to be extremely sensitive substances that quickly turn red to brownish purple in air, which is why they must always be converted into their N-acyl derivatives as quickly as possible for isolation. These are well-crystallized, stable compounds.

Since lysergic acid and isolysergic acid differ only in the steric position of the carboxyl group at carbon atom 8 and the second center of asymmetry at carbon atom 5 is removed during the transition from Ia to IIa, the use of derivatives of optically active lysergic or isolysergic acid as rearrangement products was expected to produce compounds that would behave like optical antipodes to each other. In fact, lysergol and isolysergol, in which no lysergic acid-isolysergic acid isomerization can occur prior to the indole-benzindoline rearrangement, yielded two benz[c,d] indoline derivatives when treated with proton donors, which matched in all other properties with exactly opposite optical rotation.

Synthesis of the rearrangement product

The benz[c,d]indoline structure of the rearrangement products of lysergic acid was demonstrated by the synthesis of racemic ethyl ester III[139, 145] (see Formula Scheme 11). 4,5-Dihydro-9-ethoxycarbonyl-indolo[4,3-fg] quinoline (I), an intermediate product of a dihydro-nor-lysergic acid synthesis, was converted after acetylation of the indoline nitrogen into the iodomethylate and this into the chloromethylated II. Reduction with platinum as a catalyst and the addition of two moles of hydrogen gave the tetrahydro compound III, which was identical to the rearrangement product of racemic lysergic acid ethyl ester of natural origin.

Formula Scheme 11

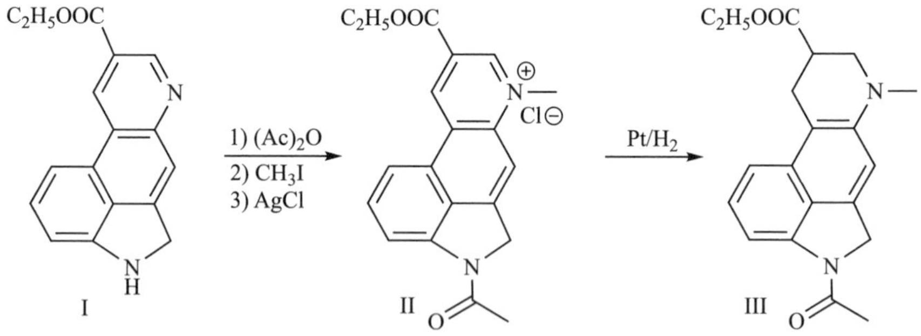

The synthetically accessible benzindoline isomer of lysergic acid was of particular interest because it differs from lysergic acid only in the position of two double bonds, the displacement of which could lead to synthetic lysergic acid. In fact, the isomerization proceeds from the lysergic acid to the benz[c,d]indoline system irreversibly, which is due to the fact that the latter is strongly energetically favored. For the benz[c,d]indoline derivative, a resonance energy of about 93 kcal is calculated, compared to about 73 kcal for the lysergic acid system conjugated with a double bond.

3. Structural elucidation of ergobasine

The relatively simple structure of ergobasine was elucidated by W. A. Jacobs and L. C. Craig[84], who showed that during alkaline hydrolysis lysergic acid and L-(+)-2-aminopropanol (= L-alaninol) are formed. Since no free amino group and no free carboxyl were detectable in ergobasine, the two components had to be linked to each other in an acid-amide manner.

The structure of ergobasine and ergobasinine determined by degradation was soon confirmed by the synthesis of the alkaloid and its stereo-isomeric forms from lysergic acid and 2-aminopropanol[89, 157].

4. Synthesis of ergobasine and its stereoisomers

The partial synthesis of ergobasine, the acid-amide-like linkage of d-lysergic acid with L-(+)-2-aminopropanol, was only feasible in a roundabout way according to the first methods described by A. Stoll and A. Hofmann[89]. These authors used the Curtius azide method for the condensation of the acid with the amino alcohol. When the hydrazide is prepared from the d-lysergic acid methyl ester or directly from the alkaloids of the ergotamine or ergotoxine group by heating with hydrazine, racemization of the lysergic acid residue and simultaneous isomerization occur, and the racemic isolysergic acid hydrazide was obtained as the reaction product[116]. This easy racemization by hydrazine is remarkable because the lysergic acid residue does not lose its optical activity when boiled with strong sodium or potassium hydroxide solution. Thus, to get to ergobasine, racemate resolution and re-isomerization was necessary.

In the first variant of the process, the racemic isolysergic acid hydrazide was converted into the azide and this was reacted with L-(+)-2-aminopropanol (L-alaninol)[89]. The resulting mixture of d-isolysergic acid L-propanolamide-(2) and l-isolysergic acid L-propanolamide-(2) could be separated chromatographically on the aluminum oxide column. The synthetic d-isolysergic acid L-propanolamide-(2) was identical to natural ergobasinine. This could be rearranged by boiling with alcoholic phosphoric acid into *d*-lysergic acid L-propanolamide-(2), which matched the natural alkaloid ergobasine in all properties.

More rational and feasible on a technical scale is the second variant of the process[157], which is characterized by the fact that the racemic isolysergic acid hydrazide resolves into its optical antipodes using the specially for this cleavage synthesized di-(*p*-toluyl)-tartaric acid[117].

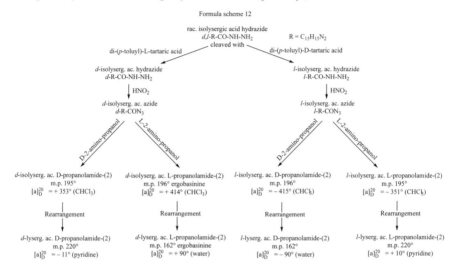

Formula scheme 12

The linkage of the optically active isolysergic acid hydrazides thus obtained via the corresponding azides with L- and D-2-aminopropanol led to the four theoretically possible stereoisomeric forms of ergobasi-nine. The rearrangement of the isolysergic acid compounds into the lyser-gic acid forms can be carried out by boiling with alcoholic phosphoric acid, or even more advantageously by standing with alcoholic potassium hydroxide solution at room temperature. An equilibrium of the isomeric forms is always formed, from which the lysergic acid isomers can be separated by fractional crystallization in the form of the free bases or suitable salts. In Formula Scheme 12, the synthesis pathway of the eight stereoisomeric forms of ergobasine is shown schematically, of which only two, *d*-lysergic acid L-propanolamide-(2) (=ergobasine) and *d*-isolysergic acid L-propanolamide-(2) (=ergobasinine), occur in nature.

Pharmacological testing of the isomers of ergobasine on the rabbit uterus *in situ* showed that the oxytocic activity is very stereospecific. Of the eight isomers differing only by the spatial arrangement at the three centers of asymmetry, only two, the natural ergobasine and the *d*-lysergic acid D-propanolamide-(2), showed the high uterine contractile activity. All isomers with *l*-lysergic acid or isolysergic acid configuration were found to be practically ineffective[157]. The oxytocic efficacy of ergobasine could be further increased by extending the amino alcohol side chain. The *d*-lysergic acid (+)-butanolamide-(2), which is listed in Section VII/1 under the synthetic derivatives of lysergic acid, has found its way into obstetrics as a uterotonicum and hemostatic agent.

5. Structural elucidation of the peptide alkaloids

The most important ergot alkaloids in terms of occurrence and practical use are peptide-like derivatives of lysergic acid. The alkaloids of this group are all structured following the same pattern. On complete hydrolysis they give lysergic acid, two amino acids (one of which is always proline), an α-keto acid, and an equivalent of ammonia[61, 69, 86, 87, 159, 160]. The second variable amino acid found is L-phenylalanine, L-leucine, or L-valine and, as α-keto acids, pyruvic acid, dimethylpyruvic acid, or α-ketobutyric acid[80].

The amino acid proline, which is contained in all alkaloids, was obtained in the D-form during acid hydrolysis, whereas under mild alkaline cleavage conditions, it was obtained as L-proline.

Based on the difference of the α-keto acid, the ergot alkaloids of the peptide type can be divided into three subgroups, within which the individual alkaloids differ only by the second, variable amino acid. Table 12 provides an overview of the cleavage products of complete hydrolysis and represents a kind of periodic system of peptide parent ergot alkaloids.

Table 12

The cleavage products of hydrolytic degradation of
ergot alkaloids of the peptide type

d-Lysergic acid (*d*-isolysergic acid)	NH₃ L-Proline	Pyruvic acid	L-Phenylalanine:	Ergotamine (Ergotaminine)	Ergotamine group
			L-Leucine:	Ergosine (Ergosinine)	
			L-Valine:	[Ergovaline (Ergovalinine)]*	
		Dimethylpyruvic acid	L-Phenylalanine:	Ergocristine (Ergocristinine)	Ergotoxine group
			L-Leucine:	Ergocryptine (Ergocryptinine)	
			L-Valine:	Ergocornine (Ergocorninine)	
		α-Ketobutyric acid	L-Phenylalanine:	Ergostine (Ergostinine)	Ergoxine group
			L-Leucine:	– – – – – – –	
			L-Valine:	– – – – – – –	

*Not yet found in nature. Synthetically prepared.

The two alkaloid pairs that yield pyruvic acid are called the ergotamine group after their prominent representative. The three alkaloid pairs
that give dimethylpyruvic acid on hydrolysis are grouped together as alkaloids of the ergotoxine group. This designation arose because a mixture
of ergocristine, ergocryptine, and ergocornine, in which the components
were contained in varying proportions, was for decades thought to be a
uniform alkaloid, which had been named ergotoxine by its discoverers
G. *Barger* and F. H. *Carr* in 1906[63]. It was not until 1943 that A. *Stoll*
and A. *Hofmann* succeeded in resolving ergotoxine into its three uniform
components with the help of di-(*p*-toluyl)-L-tartaric acid as a salt-forming
component[71].

In the "periodic table" of ergot alkaloids, one member is still missing
in the ergotamine group, namely the alkaloid corresponding to ergocornine with valine as the second amino acid, which could be called
"ergovaline." Nevertheless, ergovaline has already been totally synthesized (see Section III/6b).

Although only one alkaloid pair has been found to yield the α-ketobutyric acid as a hydrolysis product, i.e., the ergostine-ergostinine pair[80], the term "ergoxine group" is introduced here, to indicate the position of this alkaloid pair in the systematics of the peptide alkaloids. Moreover, it is very likely that with the refined analytical methodology the two missing members of this group will also be found in ergot.

The designation "ergoxine group" is intended to express the structural relationships to the ergotoxine alkaloids, from which they formally differ by the omission of a CH_2 residue. Accordingly, one syllable was omitted from the name of the homologous ergotoxine alkaloids, so that ergo[cri]stine became ergostine, and ergo[to]xine became ergoxine.

Once the individual building blocks of the peptide side chain, as they occur during full hydrolysis, had been determined, it was necessary to determine their order and the nature of their mutual linkage in the peptide residue.

While proline, phenylalanine, leucine, and valine could be regarded as genuine building blocks, there were experimental findings that suggested that the α-keto acids were not present as such in the peptide side chain. Reduction experiments under conditions where a dimethylpyruvic acid residue should have changed into an α-hydroxyisovaleric acid residue gave hydrogenated alkaloids in the lysergic acid moiety, which still yielded dimethylpyruvic acid on hydrolysis. The corresponding α-hydroxy-α-amino acid was therefore assumed to be the precursor of α-keto acid as a building block in the peptide residue[86, 161]. Pyruvic acid would therefore have arisen by hydrolytic decomposition from α-hydroxyalanine, dimethylpyruvic acid from α-hydroxyvaline, and α-ketobutyric acid from α-hydroxy-α-aminobutyric acid:

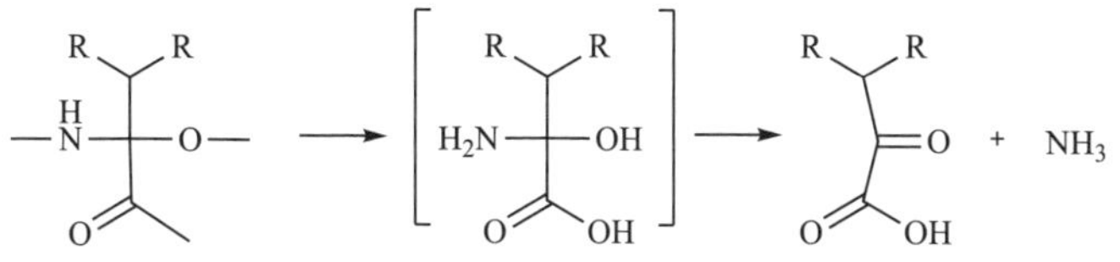

R = H or CH₃

The relative stability of the α-hydroxy-α-amino acid residue in the alkaloid molecule was explained by the fact that both the hydroxyl and the amino group were present in substituted form. Since neither a free carboxylic group nor a basic amino group was detectable in the peptide part of the alkaloids, only a cyclic arrangement of the individual building blocks could satisfy the experimental findings. Furthermore, it had to be taken into account that during the somewhat milder alkaline hydrolysis of the peptide alkaloids, mainly lysergic acid amide, instead of lysergic acid, was formed, so the lysergic acid had to be linked to the α-hydroxy-α-amino acid residue in an acid-amide manner. From these considerations, the structures A and B for the peptide alkaloids were proposed[161, 162].

The hypothetical structures A and B differ only in the sequence of amino acids. This could be determined from larger fragments obtained during the partial hydrolysis of the alkaloids.

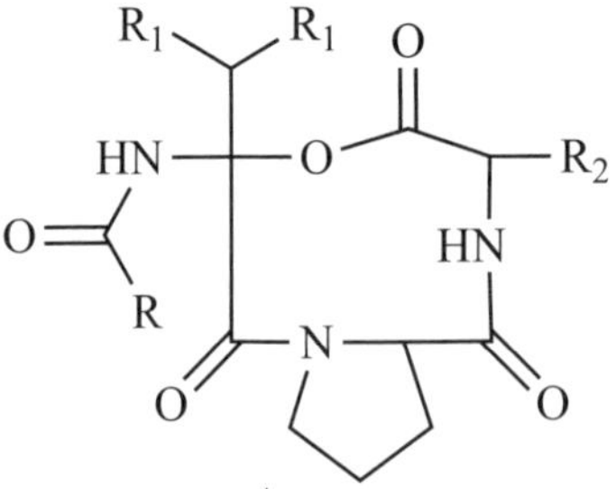

A

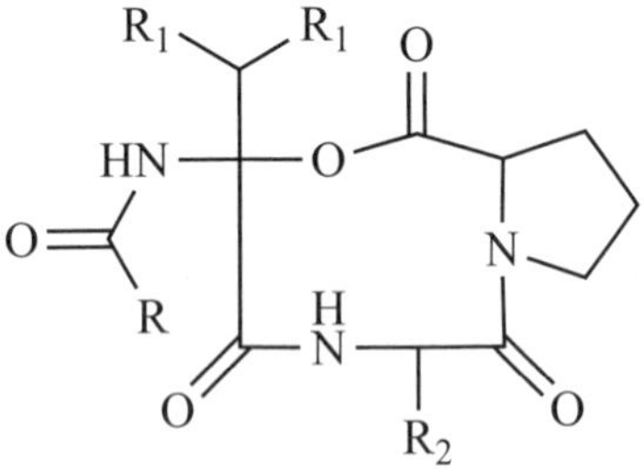

B

R = Ergolenyl residue
R_1 = H or CH_3
R_2 = Benzyl, Isobutyl or Isopropyl

Cleavage of peptide alkaloids with hydrazine

During the cleavage of peptide alkaloids or their dihydro derivatives with hydrazine, in addition to racemic isolysergic acid hydrazide or dihydrolysergic acid hydrazide, the peptide side chain could be captured as a whole. However, the hydrazine reduces one of the three building blocks that is converted to the keto acid during acidic or alkaline hydrolysis to the fatty acid. The following were obtained[163]:

from dihydroergotamine: propionyl-L-phenylalanyl-L-proline
from dihydroergocristine: isovaleryl-L-phenylalanyl-L-proline
                             and isovaleryl-L-phenylalanyl-L-
                             proline hydrazide
from dihydroergocryptine: isovaleryl-L-leucyl-L-proline
from ergocornine: isovaleryl-L-valyl-L-proline.

The structure of these tripartite peptide acids was proved by synthesis. Condensation of synthetic isovaleryl-L-valine azide with L-proline methyl ester and subsequent hydrolysis of the methyl ester moiety gave isovaleryl-L-valyl-L-proline, which was identical to the cleavage product from ergocornine. In an analogous way, propionyl-L-phenylalanyl-L-proline was synthesized, which was identified with the cleavage product from dihydroergotamine.

Thus, an acid-amide-like linkage of the carboxyl group of the precursor of the keto acid with the amino group of the variable amino acid, which in turn engages with its carboxyl group in an acid-amide-like manner on the proline, was established as a structural element of the peptide side chain.

In the hydrazine cleavage of dihydroergotamine, a considerable amount of partially racemized phenylalanyl-proline-lactam was found, and the yield of tripartite peptide acid was significantly lower than in the case of the alkaloids of the ergotoxine group, which in turn yielded only traces of the corresponding lactams. This indicates that the pyruvic acid residue, resp. its precursors in the alkaloid, when boiled with hydrazine, is cleaved significantly more readily than the corresponding precursor of dimethylpyruvic acid.

The isolation and identification of the acyl dipeptides, apart from determining the sequence of the individual building blocks in the peptide residue, also allows a statement to be made about the binding state of the proline carboxyl. The occurrence of the free carboxyl of the proline during cleavage with anhydrous hydrazine rules out an acid-amide-like linkage of this group, because the cleavage of a carboxylic acid amide with hydrazine would have to result exclusively in the corresponding hydrazide. Conversely, this finding is compatible with the presence of an ester-like or lactone-like bond, in that the formation of a free carboxyl can be formulated from such a grouping during reductive cleavage. The reduction takes place at the α-carbon atom of the postulated α-hydroxy-α-amino acid residue, since this residue indeed appears in the above-described acyl dipeptides as a fatty acid residue.

The results of the cleavage with hydrazine allowed the hypothetical structural formula A (p. 88) to be excluded. Formula B was compatible with these findings, but the α-hydroxy-α-amino acid grouping was not yet experimentally rigorously proven.

Hydrolysis with one equivalent of alcoholic alkali

By careful hydrolysis with one equivalent of aqueous alcoholic potassium hydroxide solution, the cleavage of the alkaloid molecules was accomplished in such a way that the peptide part was obtained as in the hydrazine cleavage as an acylated dipeptide, only this was not acylated with a fatty acid residue but with the corresponding α-keto acid residue. The lysergic acid part was collected as lysergic acid amide. Thus, besides the latter, the following were obtained:

from ergotamine:	pyruvoyl-L-phenylalanyl-L-proline
from ergocristine:	dimethylpyruvoyl-L-phenylalanyl-L-proline
from ergocornine:	dimethylpyruvoyl-L-valyl-L-proline.

The alkaloid molecule thus decomposes according to the following scheme, illustrated by the example of ergotamine (see p. 91).

This cleavage is therefore not a simple hydrolysis of an ordinary ester or acid amide moiety. Rather, it is based on the breakup of the building block that is directly linked to the lysergic acid residue. This labile member, hypothetically formulated as an α-hydroxy-α-amino acid moiety, is thereby converted into the pyruvic acid resp. dimethylpyruvic acid residue, while its amino group remains connected to the carboxyl group of the lysergic acid[111].

The yields of acidic peptide residue are, as in the case of cleavage with hydrazine, significantly better for the alkaloids of the ergotoxine group than of ergotamine, in which the mixed diketopiperazine of phenylalanine and proline is mainly formed with cleavage of the pyruvic acid residue.

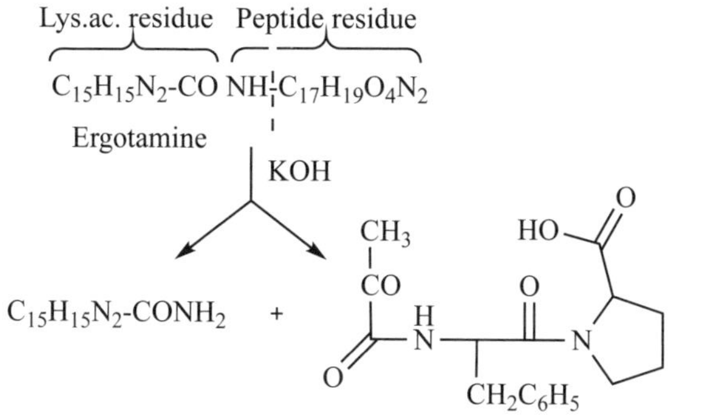

Lysergic acid amide   Pyruvoyl-L-phenylalanyl-L-proline

The reductive cleavage of peptide alkaloids with LiAlH₄

The formation of diketopiperazines under hydrolytic conditions, under which cleavage rather than the formation of a new lactam bond was to be expected, seemed difficult to reconcile with the hypothetical structure B (p. 88), and the question therefore arose as to whether the diketopiperazine ring was not already preformed in the alkaloid molecule. This consideration led to the division of the nine-membered lactam-lactone ring of formula B into a five-membered and a six-membered ring, which resulted in the initially hypothetical general structural formula C (see Formula Scheme 13) for the peptide alkaloids.

The ring closure between the nitrogen of the variable amino acid and the carbon atom of the proline carboxyl group results in a tertiary hydroxyl.

The decision between the structure B with the nine-membered lactam-lactone ring and the formula C, in which the diketopiperazine ring is preformed, could be made on the basis of the reduction products obtained with LiAlH$_4$[94]. With this reducing agent, the lactone moiety in B should have been split to form a primary and a tertiary alcohol group, whereby forming a polyamino alcohol. In fact, however, the treatment of dihydroergosine, dihydroergocristine, dihydroergocryptine, and dihydroergocornine with LiAlH$_4$ in ethylmorpholine yielded cleavage products which are assigned the general structures I, II, and III given in Formula Scheme 13. The constitutions of compounds I, II, and III could be confirmed by uniquely proceeding syntheses of the reduction products obtained from dihydroergosine and dihydroergocryptine. Since the analogous structure of all ergot alkaloids of the peptide type has been established, the knowledge gained with these bases could also be applied to the other alkaloids.

Formula Scheme 13

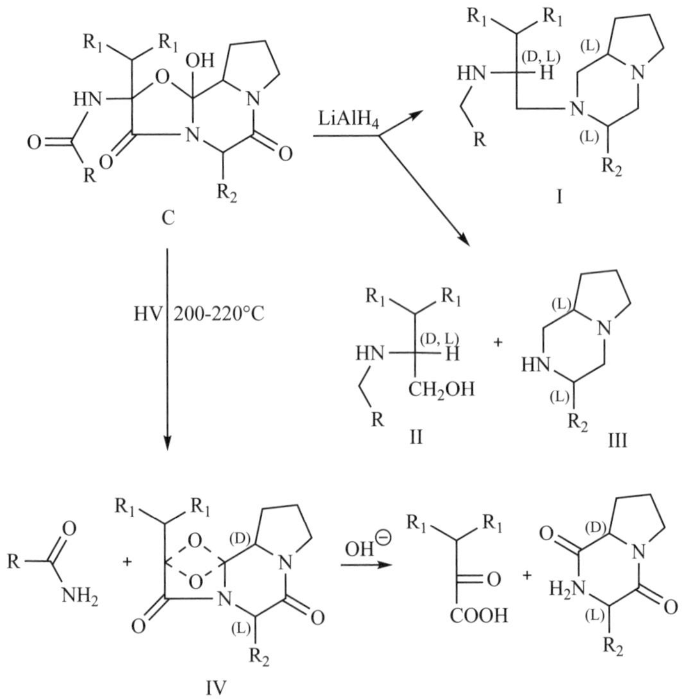

R = Ergolenyl or Ergolinyl residue
R$_1$ = H or CH$_3$
R$_2$ = Benzyl, Isobutyl or Isopropyl residue

92

The polyamines I, in which all the carbon atoms of the alkaloids used are still present, have three asymmetric carbon atoms in the reduced peptide part.

While the L-configuration of the one asymmetric center, originating from the variable amino acid, was ascertained, there was uncertainty regarding the configuration of the other two centers of asymmetry derived from the labile hydroxy-amino acid and proline. The LiAIH$_4$ reduction of the mentioned dihydro alkaloids produced two stereoisomeric polyamines I. The comparison with the synthetically prepared, sterically defined stereoisomeric forms now revealed that one isomeric polyamine I has L-configuration at the center of asymmetry of the hydroxy-amino acid residue, the other D-configuration. At the other two centers of asymmetry, both polyamines I have the L-configuration. This also provided evidence that the proline residue in the peptide alkaloids is in the L-form.

The cleavage products II contain, in addition to the reduced lysergic acid part, the amino alcohol that has arisen from the hydroxy-amino acid residue. Also here, as in polyamines I, this center of asymmetry has been racemized. This indicates that the α-carbon part of the amine alcohol is racemized. This indicates that a substituent change has taken place at the α-carbon atom as a result of the reduction, which is consistent with the formulation of a α-hydroxy-α-amino acid moiety.

The formation of piperazine III was further evidence that the diketopiperazine ring is preformed in the alkaloid molecule. Furthermore, the partial reductive cleavage of the acid amide bond between the carboxyl of the hydroxy-amino acid and the nitrogen of the variable amino acid, resulting in the cleavage fragments II and III, showed that a tertiary acid amide moiety was present, in agreement with the structural formula C. It was indeed known that tertiary but not secondary acid amides can be cleaved with LiAlH$_4$ to form aldehydes, with the aldehyde being further reduced to the alcohol by excess reagent.

The thermal cleavage of peptide alkaloids

Additional findings for the correctness of the constitutional formula C of the peptide alkaloids were obtained during thermal cleavage[94]. In high vacuum at 200–220°, the peptide alkaloids decompose rapidly, and a viscous, partially crystalline solidifying distillate was obtained in a suitable apparatus, along with a small amount of sublimate. The sublimate, which had already been obtained in 1912 by *G. Barger* and *A. J. Ewins*[164] from ergotinine and identified as dimethylpyruvic acid amide, is pyruvic acid amide, in the case of the alkaloids of the ergotamine group. The distillate consists of the diketopiperazine of proline and the variable amino acid and of a water-insoluble, well crystallizing fragment which still contains all the carbon atoms of the peptide part and which will be called "thermal cleavage product" henceforth. The distillation residue carbonizes when the natural alkaloids are used, though during thermal cleavage of the dihydro alkaloids remains as the dihydrolysergic acid amide. The alkaloid molecule breaks into two halves without losing an atom, leaving an amino group from the peptide part attached to the lysergic acid residue. For the formulation of the thermal cleavage products resulting from the peptide residues according to IV (see Formula Scheme 13), it was decisive that these compounds hydrolytically decompose at the slightest contact with alkali into pyruvic acid, or dimethylpyruvic acid, and diketopiperazine. The proline residue in diketopiperazine has a D-configuration, so inversion must have occurred. The behavior of the thermal cleavage products during catalytic reduction has not yet been completely clarified. Even under vigorous conditions, under which the free carbonyl of the pyruvic acid residue should be reduced smoothly, the pyrolysis products IV do not absorb hydrogen. This led to the unusual formulation of a 1,3-dioxacyclobutane ring, whose existence, however, seems questionable. The pyro-peptide part could later also be totally synthesized by two different methods, by cyclization of the corresponding acidic pyruvoyl, or dimethylpyruvoyl, dipeptides with acetic anhydride[165, 369] and by base-catalyzed isocyanic acid cleavage from the so-called cyclol isocyanates (see in Section III/6a).

The formation of a cleavage fragment of structure IV (see Formula Scheme 13) was difficult to deduce from formula B, but could be derived without difficulty from structural formula C. According to formula C, a free H-atom is available in the form of the tertiary OH-group, which facilitates the detachment of the lysergic acid residue as lysergic acid amide.

For the structure contained in the formula C, which was formed by cyclization with hydrogen shift from the nitrogen of an amide group to the oxygen of an adjacent lactone-carbonyl group, the term cyclol moiety was adopted. This term was introduced at the time by *D. M. Wrinch*[166, 167] for the characterization of hypothetical peptide structures in which ring formations were postulated by hydrogen shift from NH to CO groups of neighboring peptide bonds with the formation of tertiary OH-groups.

The cyclol moiety in the peptide part of the ergot alkaloids is based on the *ortho*-carboxylic acid form of proline. One of the three hydroxyls is linked lactone-like, the other lactam-like, and the third is free. The latter still exhibits very weak acidic properties and is the cause of the solubility of the peptide alkaloids in strong aqueous potassium or sodium hydroxide solution.

$$ R{-}O\underset{\underset{\displaystyle N}{|}}{\overset{\overset{\displaystyle OH}{|}}{C}}{-}R' $$

The R''—N—R''' moiety at the carbon atom of the proline carboxyl exhibits the reactions of a potential acid amide group, which explains why all oxygen is reduced out during LiAlH$_4$ reduction.

According to the investigations described above, the structure of the peptide-type ergot alkaloids was largely ascertained. In the peptide part of the ergot alkaloids, two novel structural elements, the α-hydroxy-α-aminocarboxylic acid grouping and the cyclol moiety, were found that were previously unknown in nature.

This work had also elucidated the stereochemistry at four of the six centers of asymmetry present in the alkaloid molecule, namely in the lysergic acid part, in the proline residue, and in the variable amino acid residue. Still open was the question of the configuration at the α-carbon atom of the α-hydroxy-α-amino acid residue and at the cyclol carbon atom.

The total synthesis of the peptide part and thus of the natural ergot alkaloids, which will be discussed in the next section, confirmed the constitutional formulae derived above. Furthermore, the still unanswered steric issues, the absolute configuration at the cyclol carbon atom and in the α-hydroxy-α-amino acid residue, could be thereby clarified. In anticipation of these latter derivations, the constitutional formulae of all ergot alkaloids of the peptide type known to date are listed in Formula Scheme 14.

Formula Scheme 14
General formula of ergot alkaloids of the peptide type

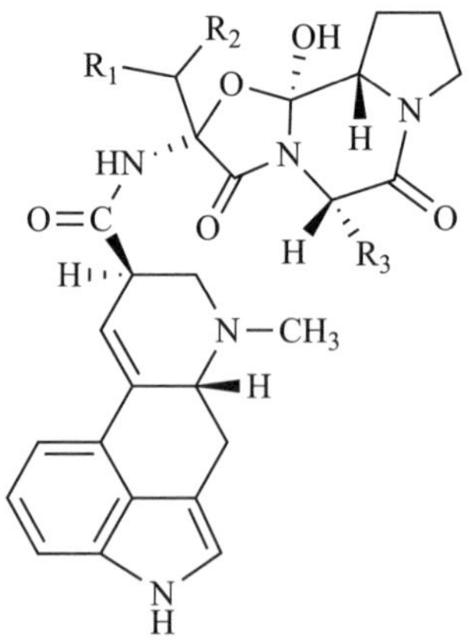

$R_1 = R_2 = H$	$R_1 = R_2 = CH_3$	$R_1 = H; R_2 = CH_3$	R_3
Ergotamine	Ergocristine	Ergostine	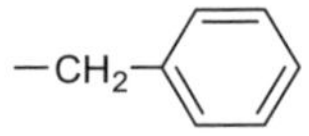 $-CH_2-$ (phenyl)
Ergosine	Ergocryptine	– – – –	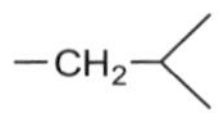 $-CH_2-$ (isopropyl)
[Ergovaline]	Ergocornine	– – – –	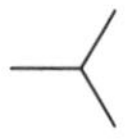 (isopropyl)

6. Synthesis of alkaloids of the ergotamine group

The difficulties in the synthesis of the peptide part of ergot alkaloids consisted in the formation of the very labile α-hydroxy-α-aminocarboxylic acid group and in the formation of the structural element known as cyclol. For these two novel moieties interlinked together in the peptide part existed no production methods yet. Extensive studies on the synthetic possibilities of α-hydroxy-α-amino acid derivatives regarding the synthesis of the peptide part of ergot alkaloids were carried out by *M. M. Shemyakin* and coworkers[168, 169]. They succeeded in incorporating such derivatives into di- and tripeptides, but cyclol formation was not possible due to the lability of the α-hydroxy-α-amino acid moiety.

The reverse path, namely the initial formation of the cyclol system and subsequent addition of the α-amino acid grouping, led *A. Hofmann, A. J. Frey*, and *H. Ott*[100] to the goal. They had found that cyclol formation occurs spontaneously when certain structural conditions are met and that cyclolization considerably stabilizes the molecule. The amino group could then be introduced in the α-position on the stable cyclol system with incorporated α-hydroxycarboxylic acid residue by conventional methods.

a) Synthesis of ergotamine

The course of the first synthesis, which led to the peptide part of ergotamine, can be seen in the Formula Scheme 15 below.

L-phenylalanyl-L-proline-lactam (II) was reacted with methyl-benzyloxymalonic acid half-ester chloride (I) in pyridine. This malonic acid derivative was obtained by reacting bromomethyl malonic acid diethyl ester with sodium benzoate to form methyl-benzyloxymalonic acid ester, which was hydrolyzed to half-ester and converted to acid chloride with thionyl chloride. The very labile acylated diketopiperazine (III) resulting from this reaction was immediately treated with Pd-hydrogen to split off the benzyl group. The resulting compound with a free hydroxyl group (IV) spontaneously cyclized to the stable cyclol form (V). The two stereoisomeric forms, formed by the reaction of the racemic malonic acid derivative with the optically active diketopiperazine, could be separated in this stable form by fractional crystallization from ethyl acetate. One stereoisomer, referred to here as A, melted at 135–136°; isomer B at 202–204°.

Formula scheme 15

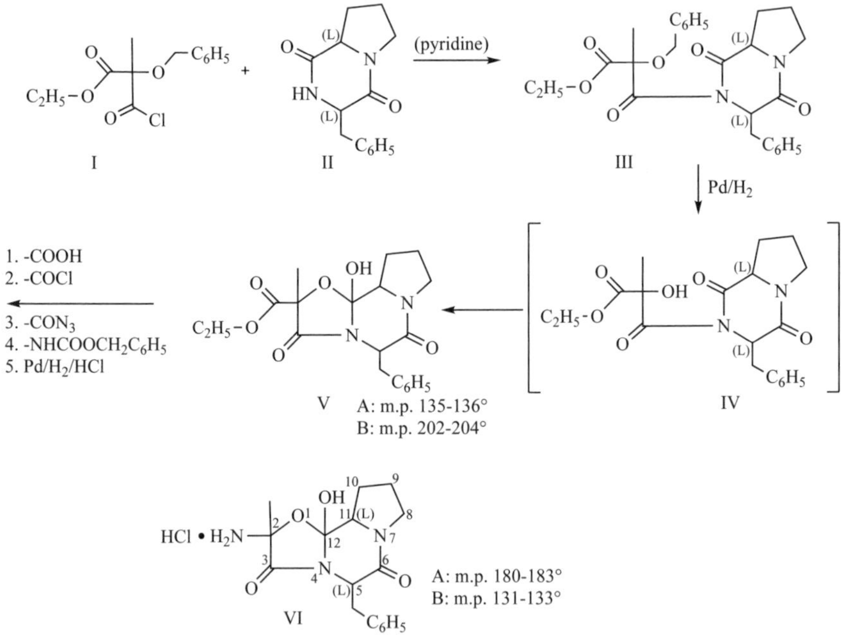

V A: m.p. 135-136°
   B: m.p. 202-204°

VI A: m.p. 180-183°
    B: m.p. 131-133°

In both isomers, the carbethoxy group was now separately converted into the amino group with the aid of a Curtius degradation. Thanks to the stability of the cyclol system, the ester could be hydrolyzed with sodium hydroxide solution to form the acid. This was converted into the acid chloride in the form of the sodium salt with oxalyl chloride, which with sodium azide yielded the acid azide. Heating with benzyl alcohol led to the benzyl urethane under Curtius rearrangement, which decomposed during hydrogenolytic cleavage into the amino cyclol (VI), which could be crystallized in the form of the stable hydrochloride. The hydrochloride of the amino cyclol from ester A melted at 180–183°, that from ester B at 131–133°. As a free base, the peptide part is extremely unstable.

Peptide hydrochloride VI A or VI B was suspended together with lysergic acid chloride hydrochloride[170] in chloroform and mixed with a chloroform solution of tributylamine at low temperature under good stirring (see Formula Scheme 16). Thereby, an acid-amide-like linkage of the two components occurs, and when using the peptide partial isomer VIA, an alkaloid was obtained which was identical in all physical, chemical, and pharmacological properties to the natural alkaloid ergotamine.

Formula Scheme 16

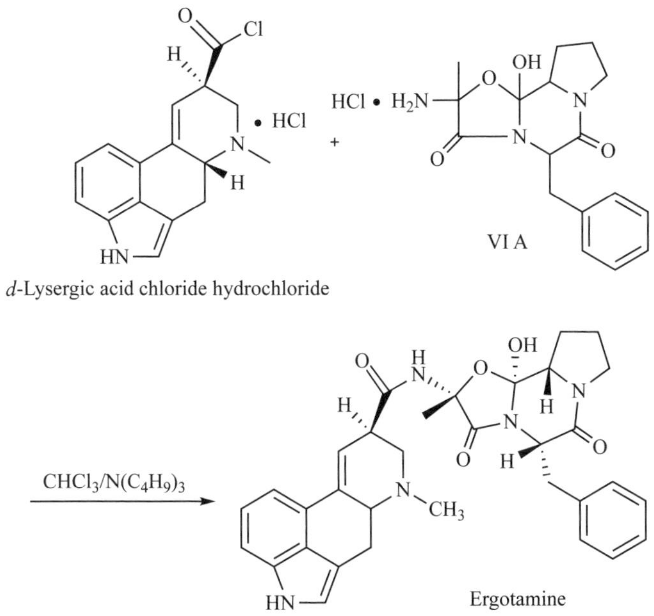

d-Lysergic acid chloride hydrochloride

VI A

CHCl₃/N(C₄H₉)₃

Ergotamine

With the peptide isomer B, an alkaloid stereoisomeric with ergotamine was obtained, which most likely differs from the latter only in the steric arrangement at the C2' of the peptide residue.

Improved synthesis of ergotamine and determination of absolute configuration at C2'

In a second, improved procedure for the synthesis of ergotamine[171], the racemic methyl-benzyloxymalonic acid half-ester (I) was separated into the optical antipodes prior to its conversion into the acid chloride[393] (see

Formula Scheme 17). With cinchonidine, the salt with the levorotatory half-ester (II B) precipitated as a poorly soluble component, whereas when anhydropilosine was used as a salt former, the dextrorotatory methyl-benzyloxymalonic acid half-ester crystallized (II A).

Formula Scheme 17

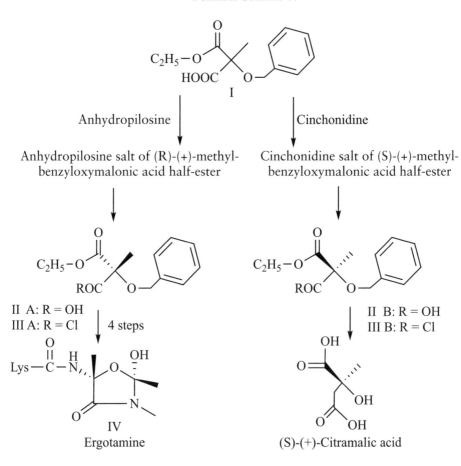

Anhydropilosine

Cinchonidine

Anhydropilosine salt of (R)-(+)-methyl-benzyloxymalonic acid half-ester

Cinchonidine salt of (S)-(+)-methyl-benzyloxymalonic acid half-ester

II A: R = OH
III A: R = Cl   4 steps

II B: R = OH
III B: R = Cl

IV
Ergotamine

(S)-(+)-Citramalic acid

The absolute configuration of the half-ester and the half-ester acid chlorides, as expressed in the stereoformulae II A and B, resp. III A and B, was determined by linking III B with the (S)-(+)-citramalic acid, whose absolute configuration had been determined by *D. Arigoni* and coworkers[285].

The use of the (S)-(-)methyl-benzyloxymalonic acid half-ester acid chloride (III A) for the synthesis of the peptide part led to the sterically uniform cyclol-carboxylic acid ester A of m.p. 135–136° (see Formula Scheme 15), which had been obtained by the first variant of the peptide moiety synthesis by fractional crystallization of the diastereomer mixture, and whose further conversion and linkage with lysergic acid provided ergotamine. This not only provided a preparatively advantageous procedure for the synthesis of ergotamine, but also elucidated the stereochemistry at the α-carbon atom of the α-hydroxy-α-aminocarboxylic acid residue as represented by stereoformula IV. Accordingly, this residue in ergotamine has the absolute configuration of α-hydroxy-L-alanine.

Thus, the absolute configuration of five of the total six asymmetry centers of the alkaloid molecule was determined.

Determination of the absolute configuration at C12'

The configuration of the cyclol hydroxyl at carbon atom 12' still remained open. This could be deduced from the reactions and measurement results shown in Formula Scheme 18.

The ring closure to the cyclol, in which a new center of asymmetry is formed at C12', is stereospecific, i.e., only one of the two theoretically possible epimeric forms is formed. The configuration of the cyclol-OH is the same in the A and B series, because the decarboxylation of cyclol carboxylic acid A (ergotamine series) and cyclol carboxylic acid B produced an identical decarboxylation product V. The position of the cyclol-OH relative to the carboxyl group in the cyclol carboxylic acids, the configuration of which was derived in the previous section, must have an effect on the acidity of these acids, depending on whether hydrogen bonds can form or not. It is known that the acidity of acids with hydrogen bonds is increased, while that of the hydroxyl groups forming hydrogen bonds is decreased. The pK values of the two isomeric cyclol carboxylic acids, which are given under the corresponding formulae, indicate that hydrogen bridges are present in the A series, i.e., *cis*-position of the cyclol-OH to the carboxyl and *trans*-arrangement in the B series.

Formula Scheme 18

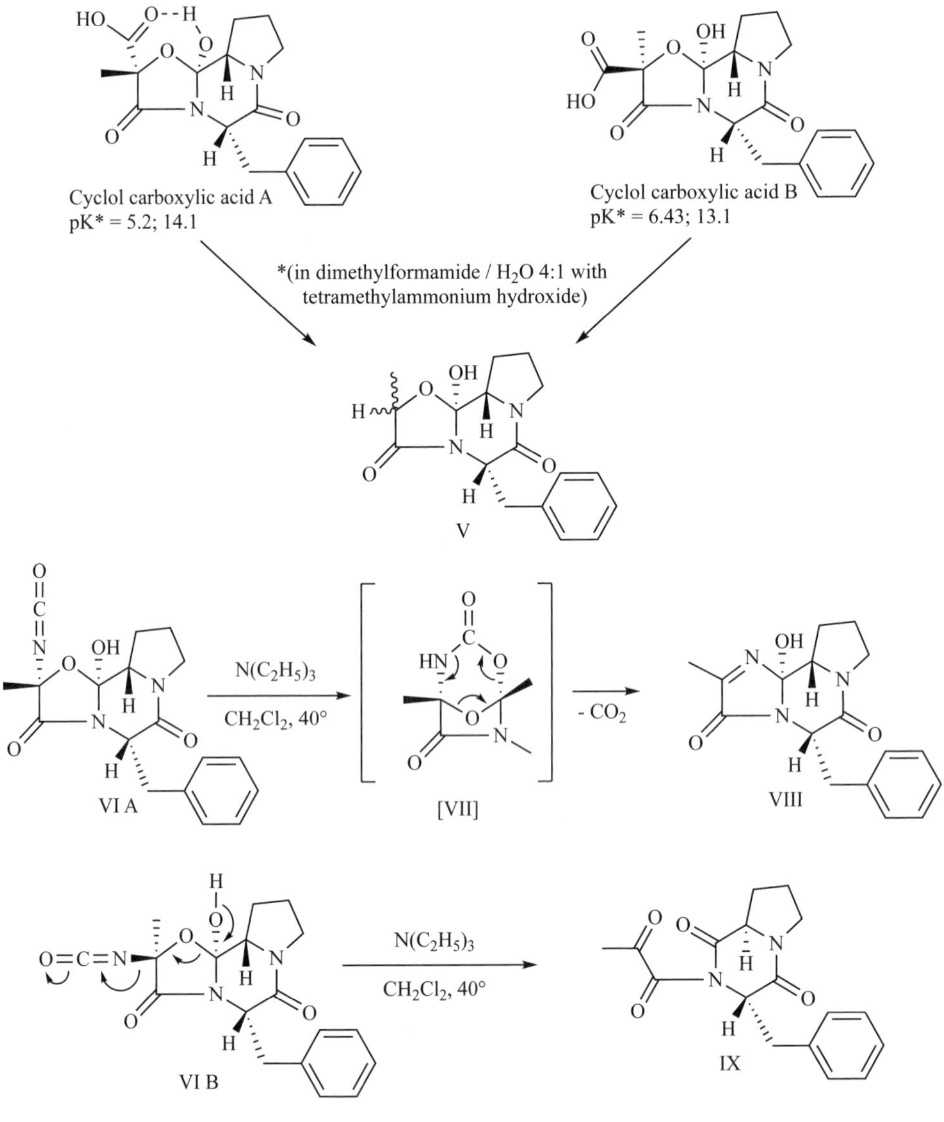

Cyclol carboxylic acid A
pK* = 5.2; 14.1

Cyclol carboxylic acid B
pK* = 6.43; 13.1

*(in dimethylformamide / H₂O 4:1 with tetramethylammonium hydroxide)

V

VI A   [VII]   VIII

$N(C_2H_5)_3$
CH_2Cl_2, 40°

$- CO_2$

VI B   IX

$N(C_2H_5)_3$
CH_2Cl_2, 40°

The same result was obtained on the basis of the different course of the base-catalyzed conversion of the cyclol carboxylic acids corresponding isocyanates VI A and VI B. From the isocyanate of the A series, the imidazolinone derivative VIII is formed, the formation of which can be understood by the intermediate intramolecular urethane VII, which requires the *cis*-position of OH- and NCO- groups. In contrast, the isocyanate VI B undergoes a transformation into the pyruvoyl-diketopiperazine IX based on a *trans*-elimination. IX was already obtained during the thermal cleavage of ergotamine (see Section C, III/5). In parallel with the cleavage, in both cases an isomerization of the L-proline residue into the D-form takes place.

From the analogous structure of all alkaloids of the peptide type, it may be concluded that they also correspond sterically with each other. Thus, the structural and steric construction of the peptide ergot alkaloids can be represented by the general formula as shown at the end of Section C, III/5 (see Formula Diagram 14).

b) Synthesis of the ergosine and the valine analog of
the ergotamine group

Ergosine, which differs in its structure from ergotamine only by the replacement of the L-phenylalanine residue by the L-leucine residue, could be prepared by the same process as ergotamine. The alkaloid corresponding to ergocornine in the ergotamine group with L-valine as the variable amino acid, which has not yet been found in nature, has also been produced synthetically. Thus, this gap in the systematics of the natural peptide alkaloids (see Table 12) has been filled synthetically.

The synthetic pathway for these two alkaloids can be seen in Formula Scheme 19[203].

The acid chloride of the (R)-(+)-methyl-benzyloxymalonic acid half-ester (IIIA), which was already used for the synthesis of the ergotamine peptide moiety, was condensed in pyridine solution with the diketopiperazine of L-proline and L-leucine (X a), resp. L-proline and L-valine (X b), to the very easily hydrolysable acylation product XI a, resp. XI b. After the hydrogenolytic cleavage of the benzyl group, cyclolization to the corresponding sterically uniform cyclol carboxylic acid ester XII occurred spontaneously. Cyclol formation was, as in the case of ergotamine synthesis, stereospecific, in that only one of the two at C12' epimeric cyclols were formed. The Curtius degradation of the azides XII 4a, resp. XII 4b, led to the corresponding amino cyclols, which could be captured in crystallized form in the form of the hydrochlorides XIII 2a and XIII 2b.

Condensation of XIII 2a with *d*-lysergic acid chloride hydrochloride yielded an alkaloid identical to the natural ergot alkaloid ergosine.

Acid amide-like linkage of XIII 2b with *d*-lysergic acid yielded the to ergocornine corresponding third alkaloid of the ergotamine series; if it were subsequently found in ergot, and then a trivial name justified, it could be called "ergovaline."

Formula Scheme 19

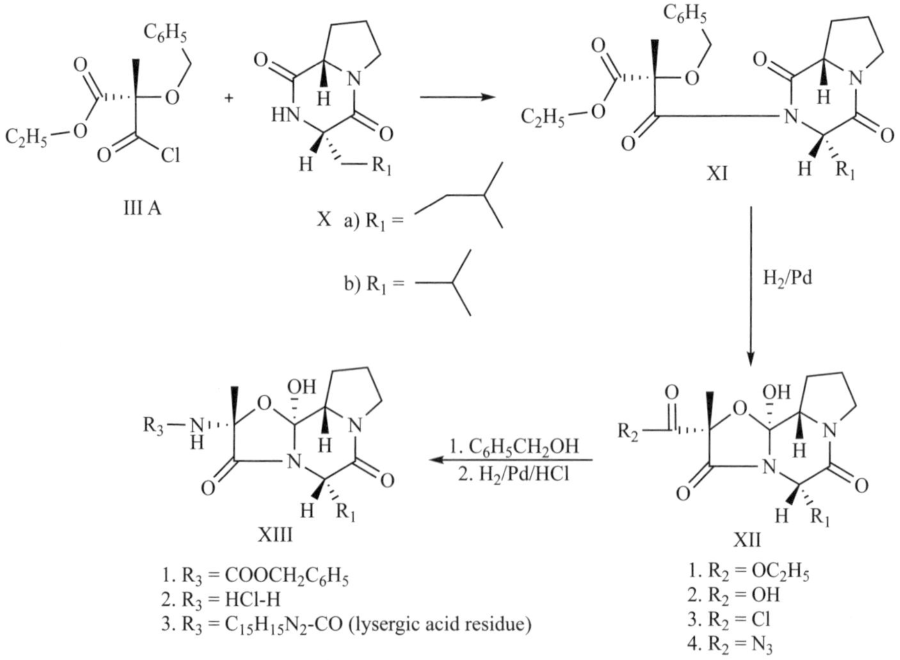
C6H5
O
C2H5—O
O
Cl
+
HN
H
N
O
H
—R1
III A
X a) R1 =
b) R1 =
C2H5—O
XI
H R1
H2/Pd
R3—N
H
OH
XIII
1. C6H5CH2OH
2. H2/Pd/HCl
R2
XII
1. R3 = COOCH2C6H5
2. R3 = HCl-H
3. R3 = C15H15N2-CO (lysergic acid residue)
1. R2 = OC2H5
2. R2 = OH
3. R2 = Cl
4. R2 = N3

7. Synthesis of ergostine

The synthesis worked out on the example of ergotamine offers possibilities to vary the peptide part of ergot alkaloids in many ways while retaining the cyclol structure characteristic of this alkaloid group. Besides the replacement of the amino acids L-proline, L-phenylalanine, L-leucine and L-valine contained in the natural alkaloids by other α-amino acids, it is also possible to incorporate other α-hydroxy-α-amino acids into the peptide part of the molecule, instead of the α-hydroxy-L-alanine residue, which characterizes the alkaloids of the ergotamine group, or the α-hydroxyvaline residue of the alkaloids of the ergotoxine group.

As a first example of such a variation, the α-hydroxy-L-alanine residue was replaced by the α-hydroxy-α-aminobutyric acid residue in the molecule of ergotamine[80, 394].

This led to an alkaloid that is structurally intermediate between ergotamine and ergocristine in that the methyl group at C2' of ergotamine, resp. the isopropyl group of ergocristine, is replaced by an ethyl group.

Almost simultaneously with the synthetic production, W. *Schlientz* et al.[80] isolated this alkaloid, called ergostine, from ergot. The structure of ergostine was essentially elucidated by its synthesis. This can be seen in the Formula Scheme 20, entirely following the scheme of the ergotamine synthesis.

Formula Scheme 20

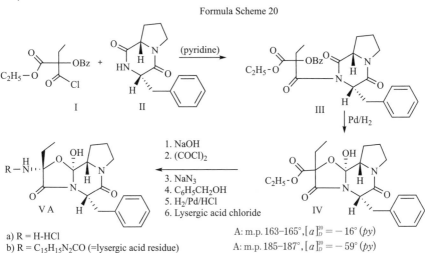

a) R = H-HCl
b) R = $C_{15}H_{15}N_2CO$ (=lysergic acid residue)

A: m.p. 163–165°, $[a]_D^{20} = -16°$ (*py*)
A: m.p. 185–187°, $[a]_D^{20} = -59°$ (*py*)

Ethyl-benzyloxymalonic acid half-ester chloride (I), which was prepared from bromoethyl malonic acid diethyl ester in the conventional way, was condensed in pyridine solution with L-phenylalanyl-L-proline lactam (II).

The cyclolization after debenzylation of III was again stereospecific, because the cyclol carboxylic acid ester IV consisted only of the two at C2' stereoisomeric components A and B, which could be separated chromatographically. The conversion of component A via a Curtius rearrangement to amino cyclol hydrochloride VA a) and linkage with *d*-lysergic acid yielded a compound that was identical to the natural ergot alkaloid ergostine. In analogy to the steric ratios determined for ergotamine, ergostine probably has the configuration given in formula VA b).

IV. The group of clavine alkaloids

The first representatives of this second main group of natural ergot alkaloids, which differ from the classical lysergic acid alkaloids in that the carboxyl group of the lysergic acid is reduced to the hydroxymethyl or methyl group, were discovered only in 1951 by *M. Abe* and coworkers in Japan in the ergot of Far Eastern grass species. These were agroclavine from the ergot of *Agropyrum semicostatum* and elymoclavine from *Elymus mollis* ergot. All other alkaloids of the same structural type isolated later were given the suffix -clavine in order to express their structural affiliation.

A particularly productive source of clavine-type alkaloids has been found in the ergot of the African foxtail millet, *Pennisetum typhoideum*. Recently, with the aid of the new, more powerful separation and detection methods, such as paper and thin-layer chromatography, clavine alkaloids have also been detected in rye ergot of various origins. Furthermore, *A. Hofmann* and *H. Tscherter* found in 1960 as a great surprise the occurrence of ergot alkaloids, of the lysergic acid and of the clavine type, for the first time in higher plants, namely in genera from the family of the morning glories (*Convolvulaceae*)[78, 112]. Another, albeit less surprising expansion of the occurrence of ergot alkaloids, was the discovery of already known and new clavine alkaloids in *Aspergillus* species (*Fungi imperfecti*) and in *Phycomyces* and *Rhizopus* species (*Phycomycetes*) by *J. F. Spilsbury* and *S. Wilkinson* in 1961[79]. This has broken the monopoly position of the genus *Claviceps* as producers of ergot alkaloids.

Table 13
The natural alkaloids of the clavine group

Name	Empirical formula	$[\alpha]_D$ (in pyridine)	First isolated by
Ergolene-(8)-derivatives			
Agroclavine	$C_{16}H_{18}N_2$	$-182°$	M. Abe (1951)
Elymoclavine	$C_{16}H_{18}ON_2$	$-152°$	M. Abe, T. Yamano, Y. Kozu and M. Kusumoto (1952)
Molliclavine	$C_{16}H_{18}O_2N_2$	$+30°$	M. Abe and S. Yamatodani (1954)
Ergolene-(9)-derivatives			
Lysergine	$C_{16}H_{18}N_2$	$+65°$	M. Abe, S. Yamatodani, T. Yamano and M. Kusumoto (1960)
Lysergol	$C_{16}H_{18}ON_2$	$+54°$	A. Hofmann and H. Tscherter (1960); M. Abe, S. Yamatodani, T. Yamano and M. Kusumoto (1960)
Lysergene	$C_{16}H_{16}N_2$	$+504°$	M. Abe, S. Yamatodani, T. Yamano and M. Kusumoto (1960)
Setoclavine	$C_{16}H_{18}ON_2$	$+174°$	} A. Hofmann, R. Brunner, H. Kobel
Isosetoclavine	$C_{16}H_{18}ON_3$	$+107°$	} and A. Brack (1957)
Penniclavine	$C_{16}H_{18}O_2N_2$	$+151°$	A. Stall, A. Brack, H. Kobel, A. Hofmann and R. Brunner (1954)
Isopenniclavine	$C_{16}H_{18}O_2N_3$	$+146°$	A. Hofmann, R. Brunner, H. Kobel and A. Brack (1957)
Ergoline-derivatives			
Festuclavine	$C_{16}H_{20}N_2$	$-110°$	M. Abe and S. Yamatodani (1954)
Pyroclavine	$C_{16}H_{20}N_3$	$-90°$	} M. Abe, S. Yamatodani, T. Yamano and
Costaclavine	$C_{16}H_{20}N_4$	$+44°$	} M. Kusumoto (1956)
Fumigaclavine A	$C_{18}H_{22}O_2N_2$	$-57°*$	} J. F. Spilsbury and S. Wilkinson (1961)
Fumigaclavine B	$C_{18}H_{22}O_2N_3$	$-113°**$	
Chanoclavine	$C_{16}H_{20}ON_2$	$-240°$	A. Hofmann, R. Brunner, H. Kobel and A. Brack (1957)

*Hydrochloride $[\alpha]_{5461}^{22}$ (in methanol). ** $[\alpha]_{5461}^{22}$

Table 13 above lists the clavine-type alkaloids described to date. They are divided into groups on the basis of structural features. In the same order, namely ergolene-(8)-, ergolene-(9)-, ergoline-, chano-derivatives, the individual alkaloids are then characterized by their most important data.

1. The individual alkaloids of the clavine group

Agroclavine
$C_{16}H_{18}N_2$ (m.w. 238.3)

Agroclavine was first found in ergot of *Agropyrum semicostatum* Nees and *A. ciliare* Fr. and in saprophytic cultures of this fungus[25]. Later it was found as one of the main alkaloids in the ergot of *Pennisetum typhoideum* Rich. and in the saprophytic cultures prepared from it[17].

Agroclavine has no pharmacological properties that could have led to medicinal use[173].

Agroclavine crystallizes from acetone in colorless needles, m.p. 205–206° (dec.)[17, 25, 72]. It is easily soluble in alcohol, chloroform, and pyridine, moderately soluble in benzene or ether, and very sparingly soluble in water. It forms salts readily soluble in water[172]. It can be sublimed at 110–130° in high vacuum without decomposition.

$$[\alpha]_D^{20} = -155° \ (c = 0.9 \ in \ chloroform)$$
$$[\alpha]_D^{20} = -182° \ (c = 0.5 \ in \ pyridine)$$
$$pK^* = 6.8 \ (in \ 80\% \ aq. \ methyl \ cellosolve)$$

In the Keller color reaction, agroclavine gives a violet-blue coloration. The UV spectrum is characterized by the following maxima: $\log \varepsilon_{max}$ = 4.47 at 225 mμ; $\log \varepsilon_{max}$ = 3.88 at 284 mμ; $\log \varepsilon_{max}$ = 3.81 at 293 mμ. IR spectrum see Section V, 3b, Figure 15r.

Elymoclavine
$C_{16}H_{18}ON_2$ (m.w. 254.3)

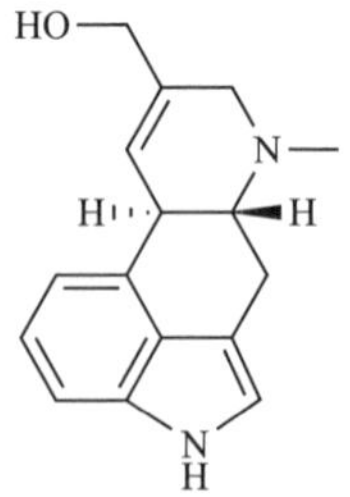

Elymoclavine received its name after the grass *Elymus mollis* Tri, native to Japan, Sakhalin, and the Kuril Islands. This alkaloid was first found in ergot growing on this grass and in saprophytic cultures of this fungus[27]. Elymoclavine, along with agroclavine, is the main alkaloid in the ergot of the tropical foxtail millet *Pennisetum typhoideum* Rieb. It can be isolated in rich yields from saprophytic cultures of this fungus[17]. It is also contained in the seeds of certain species of morning glories[78].

The pharmacological property of elymoclavine is claimed to be an excitatory effect due to stimulation of sympathetic centers[173].

Elymoclavine is sparingly soluble in the usual organic solvents. Only in pyridine does it show considerable solubility. From methanol, in which it dissolves in 70 parts at boiling heat, elymoclavine crystallizes in crystalline solvent free prisms which melt at 245–249° with decomposition[17].

$$[\alpha]_D^{20} = -111° \, (c = 0.1 \text{ in ethanol})$$
$$pK^* = 6.8 \, (\text{in 80\% aq. methyl cellosolve})$$

With Keller's reagent, elymoclavine gives a violet-blue coloration.

UV spectrum: $\log \varepsilon_{max}$ = 4.31 at 227 mμ; $\log \varepsilon_{max}$ = 3.84 at 283 mμ; $\log \varepsilon_{max}$ = 3.76 at 293 mμ.

IR spectrum. see Section V, 3b, Figure 15s.

Molliclavine

$C_{16}H_{18}O_2N_2$ (m.w. 270.3)

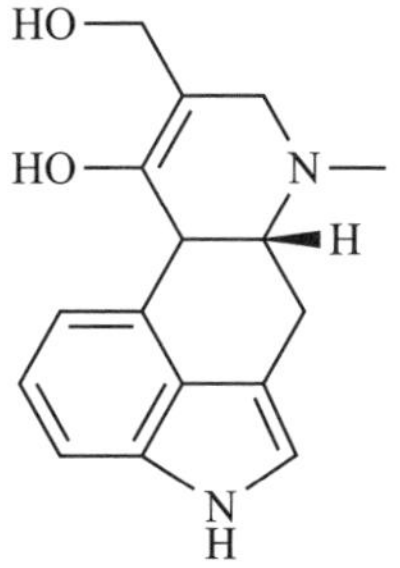

Molliclavine has been isolated from the sclerotia and saprophytic cultures of the ergot fungus of *Elymus mollis* Tri. in very small amounts[73]. It has not been detected in any other ergot species since. The structural formula reproduced above is to be regarded as provisional only.

Molliclavine crystallizes from methanol, in which it is readily soluble, or from acetone, in which it is moderately soluble, in prisms, m.p. 253° (dec.). In benzene or chloroform the alkaloid is very sparingly soluble; however, it shows considerable solubility in water.

$$[\alpha]_D^{17} = +30°; [\alpha]_{5461}^{17} = +42° \, (c = 0.2 \; in \; pyridine)$$

In the Keller's or van Urk's color reaction, a green coloration.

The UV spectrum shows maxima at 226, 287, and 294 mμ, i.e., it is almost the same with the spectrum of agroclavine or elymoclavine. During the catalytic hydrogenation a dihydro derivative is formed, which with Keller or van Urk reagent, shows the violet-blue color characteristic of ergoline derivatives.

Lysergine

$C_{16}H_{18}N_2$ (m.w. 238.3)

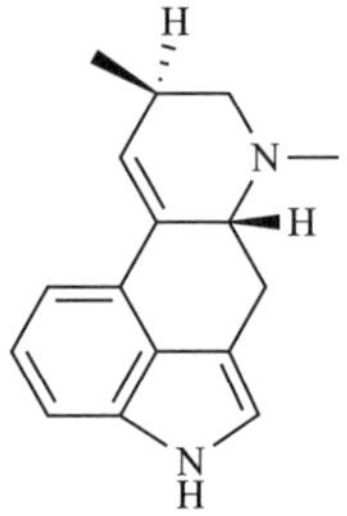

Lysergine, before it was isolated as a natural alkaloid from saprophytic cultures of the *Agropyrum* type ergot fungus[77], was prepared by treating elymoclavine or agroclavine with sodium in butanol[174] and from lysergene by catalytic reduction[175].

Lysergine crystallizes from methanol, ethanol, or from ethyl acetate, in which it is moderately soluble, in prisms, m.p. 286–289° (dec.).

$$[\alpha]_D^{20} = +65° \, (c = 0.5 \; in \; pyridine)$$

UV spectrum: same as lysergic acid

IR spectrum: see Section V, 3b, Figure 15t.

Color reactions: With Keller and van Urk reagent violet-blue coloration.

Lysergol

$C_{16}H_{18}ON_2$ (m.w. 254.3)

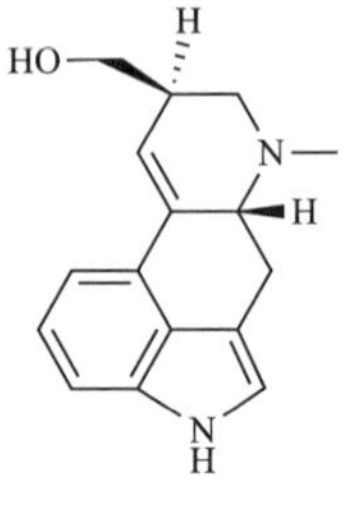

Lysergol was found in nature as late as 1960 in the seeds of a species of morning glory (*Rivea corymbosa* [L.] Hall. f., aka *Turbina corymbosa* and *Ipomea tricolor*)[78, 112] and in saprophytic cultures of an ergot fungus of the *Elymus* type[76, 77]. The compound was prepared synthetically already in 1949 by LiAlH$_4$ reduction of lysergic acid methyl ester[176].

Lysergol crystallizes from alcohol in plates and prisms, m.p. 253–255° (dec.). The alkaloid dissolves in 350 parts boiling methanol or 100 parts boiling ethanol. It is sparingly soluble in water or chloroform. It can be sublimed in high vacuum at 180° without decomposition.

$$[\alpha]_D^{20} = + 54°; [\alpha]_{5461}^{20} = + 87° \ (c = 0.3 \ in \ pyridine)$$
$$pK^* = 6.6 \ (in \ 80\% \ aq. \ methyl \ cellosolve)$$

IR spectrum: see Section V, 3b, Figure 15u.

Tartrate, needles from alcohol, m.p. 231°C (dec.).

Lysergene
C$_{16}$H$_{16}$N$_2$ (m.w. 236.3)

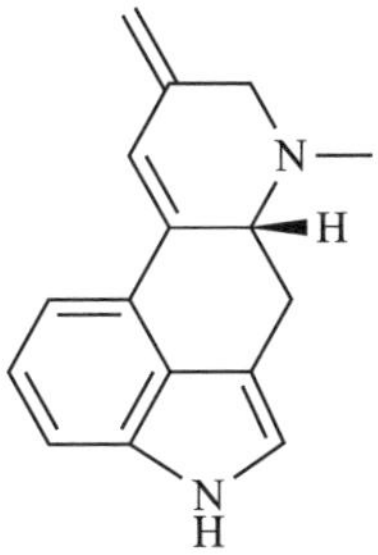

Lysergene has been isolated from saprophytic cultures of an ergot fungus of the *Elymus* type[76, 77]. Previously, it was produced from elymoclavine by chemical means[174].

Lysergene crystallizes from methanol in colorless prisms or needles, m.p. 247–249° (dec.). The compound is sparingly soluble in most organic solvents, moderately soluble in chloroform or in pyridine.

$$[\alpha]_D^{20} = + 504° \ (c = 0.4 \ in \ pyridine)$$

UV spectrum: maxima at 243, 263 and 335 mμ.

Color reactions: With Keller's reagent green coloration, which gradually fades to yellow. With conc. sulfuric acid blue coloration.

Setoclavine
$C_{16}H_{18}ON_2$ (m.w. 254.3)

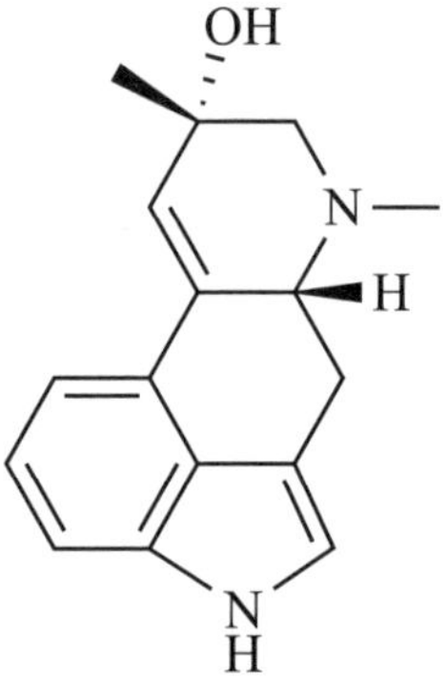

Setoclavine was first isolated from saprophytic cultures of the ergot fungus of the African foxtail millet, *Pennisetum typhoideum* Rich.[18]. The same alkaloid was obtained from sclerotia and saprophytic cultures of ergot fungus from grasses growing in Japan (*Elymus mollis, Agropyrum semicostatum, Trisetum bifidum, Festuca rubra*) in a form that was not yet completely pure and with accordingly deviating data described as triseclavine[177].

Setoclavine crystallizes from acetone or methanol in solid prisms free from crystallizing solvents, m.p. 229–234° (dec.). It dissolves at boiling heat in 40 parts of the above solvents, in 50 parts ethyl acetate, in 40 parts chloroform, or in 15 parts dioxane. It is sparingly soluble in water.

$$[\alpha]_D^{20} = + 174°; [\alpha]_{5461}^{20} = + 232° \ (c = 1.1 \ in \ pyridine)$$
$$pK^* = 6.4 \ (in \ 80\% \ aq. \ methyl \ cellosolve)$$

UV spectrum: maxima at 243 mμ (log ε_{max} = 4.38) and at 313 mμ (log ε_{max} = 4.04).

IR spectrum: see Section V, 3b, Figure 15v.

In the Keller's and van Urk's color reactions, a green coloration is formed. When a trace of setoclavine is dissolved in conc. sulfuric acid, a pure blue color is formed which is stable for several hours.

Salts of setoclavine:

Hydrochloride from water or alcohol needles, which turn dark from 200° without melting up to 300°.

Nitrate from water needles, which, like hydrochloride, do not melt up to 300°.

Isosetoclavine

$C_{16}H_{18}ON_2$ (m.w. 254.3)

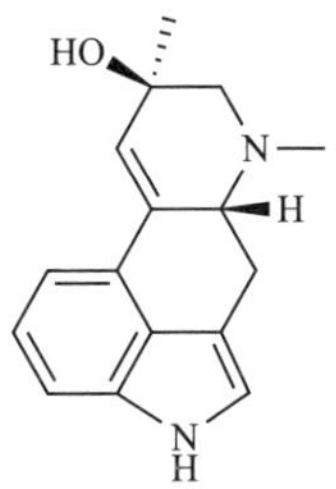

Isosetoclavine, together with setoclavine, was first isolated from the ergot fungus of *Pennisetum typhoideum* Rich.[18] and was also found in the ergot of grasses native to Japan[74].

From methanol solvent-free polyhedra, m.p. 234–237° (dec.). Dissolves at boiling heat in 70 parts methanol, 60 parts acetone, or 160 parts chloroform.

$$[\alpha]_D^{20} = +107°; [\alpha]_{5461}^{20} = +147° \, (c = 0.5 \, in \, pyridine)$$
$$pK^* = 5.9 \, (in \, 80\% \, aq. \, methyl \, cellosolve)$$

UV spectrum: maxima at 242 mμ (log ε_{max} = 4.42) and at 317 mμ (log ε_{max} = 4.10).

IR spectrum see Section V, 3b, Figure 15w.

Gives the same color reactions as setoclavine.

Hydrochloride crystallizes from methanol when diluted with acetone in rosettes, which do not melt up to 300°.

Penniclavine

$C_{16}H_{18}O_2N_2$ (m.w. 270.3)

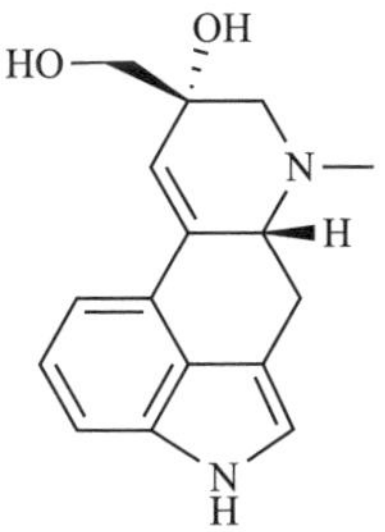

Discovered as a product of the ergot fungus of *Pennisetum typhoideum* Rich.[17], penniclavine was later also found in the ergot of Japanese wild grasses (*Agropyrum semicostatum* Nees, *Trisetum bifidum* Ohwi, *Festuca rubra* L.)[177] and also in very small amounts in rye ergot[178].

Penniclavine dissolves at boiling heat in 30 parts methanol or in 80 parts acetone and separates from these solvents in rectangular platelets containing no solvent, m.p. 222–225° (dec.).

$$[\alpha]_D^{20} = +\ 151°; [\alpha]_{5461}^{20} = +\ 201°\ (c = 0.5\ in\ pyridine)$$
$$pK^* = 6.4\ (in\ 80\%\ aq.\ methyl\ cellosolve)$$

UV spectrum: maxima at 240 mµ (log ε_{max} = 4.29) and at 313 mµ (log ε_{max} = 3.93)

IR spectrum: see Section V, 3b, Figure 15x.

Penniclavine gives a green color in the Keller and van Urk color reactions and dissolves in conc. sulfuric acid with a blue color.

Isopenniclavine

$C_{15}H_{18}O_2N_2$ (m.w. 270.3)

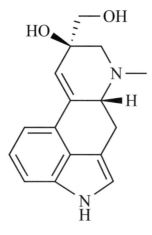

Isopenniclavine was discovered in saprophytic cultures of the ergot fungus of *Pennisetum typhoideum* Rich. among a number of other clavine alkaloids[18].

Isopenniclavine crystallizes from water, in which it dissolves to 1% at boiling heat, in hexagonal plates without crystal solvent. The alkaloid is very slightly soluble in methanol and acetone, slightly soluble in ethyl acetate, and moderately soluble in chloroform, m.p. 163–165° (dec.).

$$[\alpha]_D^{20} = + 146°; [\alpha]_{5461}^{20} = + 198° \ (c = 0.7 \ in \ pyridine)$$
$$pK^* = 5.9 \ (in \ 80\% \ aq. \ methyl \ cellosolve)$$

UV spectrum: maxima at 242 mµ (log ε_{max} = 4.31) and at 313 mµ (log ε_{max} = 3.94)

IR spectrum: see Section V, 3b, Figure 15y.

Color reactions like penniclavine.

Festuclavine

$C_{16}H_{20}N_2$ (m.w. 240.3)

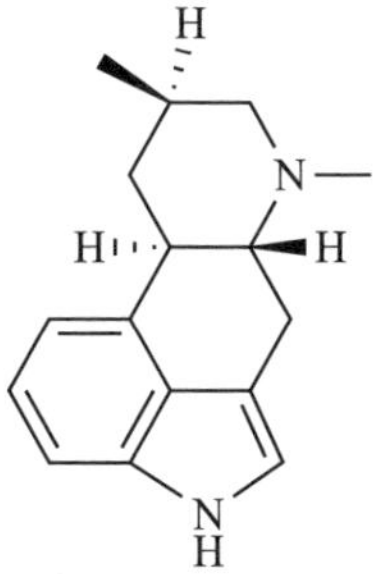

Festuclavine was first found as a product of the ergot fungus of grasses of the genus *Agropyrum* and *Phalaris*[72]. However, this alkaloid has also been isolated from cultures of *Aspergillus fumigatus* Fres. in addition to fumigaclavine A and B[79]. Synthetically, festuclavine could be obtained by catalytic hydrogenation of agroclavine[175, 179] or by reduction with sodium in butanol[174].

Festuclavine crystallizes from methanol in fine needles, m.p. 242–244° (dec.). It is readily soluble in alcohol, chloroform, and acetone and moderately soluble in benzene or ether. The alkaloid can be sublimed without decomposition.

$$[\alpha]_D^{20} = - 70°; [\alpha]_{5461}^{20} = - 83° \ (c = 0.5 \ in \ chloroform)$$
$$[\alpha]_D^{20} = - 110°; [\alpha]_{5461}^{20} = - 128° \ (c = 0.5 \ in \ pyridine)$$
$$pK^* = 6.4 \ (in \ 80\% \ aq. \ methyl \ cellosolve)$$

UV spectrum: maxima at 224 mμ (log ε_{max} = 4.54), 276 mμ (log ε_{max} = 3.81), 281 mμ (log ε_{max} = 3.84)

IR spectrum: see Section V, 3b, Figure 15z, 1.

Color reactions: With Keller and with van Urk reagent violet-blue coloration.

Pyroclavine

$C_{16}H_{20}N_2$ (m.w. 240.3)

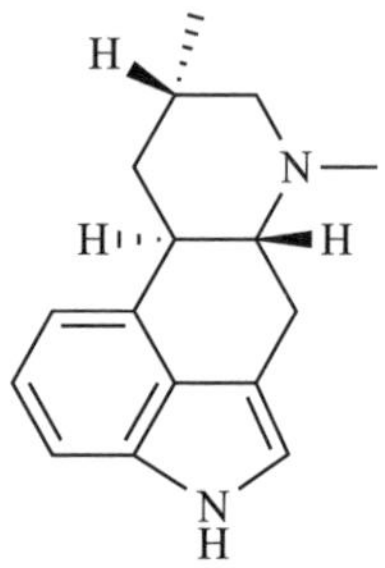

Pyroclavine was found in the mother liquors of festuclavine in the alkaloid fraction from the sclerotia and saprophytic cultures of the *Agropyrum* type of ergot fungus[74]. It could also be produced synthetically by reduction with sodium in butanol[174] or by catalytic reduction of agroclavine with palladium in alcohol[175].

Pyroclavine crystallizes from ethyl acetate, benzene, or methanol in needles, m.p. 204° (dec.). The alkaloid can be sublimed undecomposed in high vacuum.

$$[\alpha]_D^{20} = -90°; \quad [\alpha]_{5461}^{20} = -105° \ (c = 0.5 \ in \ pyridine)$$

UV spectrum shows maxima at 224, 275, 282 ane 292 mμ.

Color reactions: Like festuclavine.

Costaclavine

$C_{16}H_{20}N_2$ (m.w. 240.3)

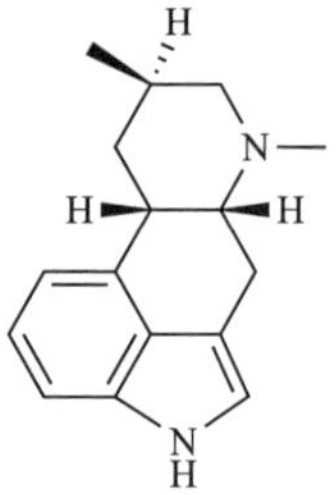

Costaclavine was isolated from saprophytic cultures of the *Agropyrum* type ergot fungus along with other clavine alkaloids[74].

Synthetically, costaclavine was obtained during the reduction of agroclavine or elymoclavine with sodium in butanol among other hydrogenation and rearrangement products[174].

Costaclavine crystallizes from ethyl acetate, acetone, methanol, or ethanol in prisms, m.p. 182° (dec.). The alkaloid is moderately soluble in acetic ester or chloroform, but readily soluble in methyl and ethyl alcohol.

$$[\alpha]_D^{20} = +44°; \ [\alpha]_{5461}^{20} = +59° \ (c = 0.2 \ in \ pyridine)$$

The UV spectrum is identical to that of pyroclavine.

Fumigaclavine A
$$C_{18}H_{22}O_2N_2 \ (m.w. \ 298.4)$$

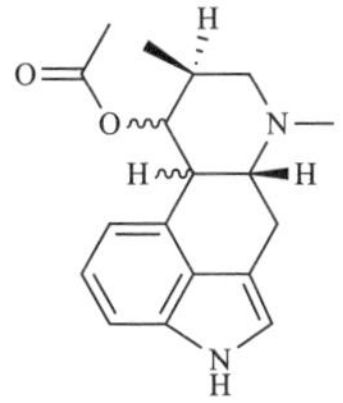

Fumigaclavine A has been isolated as the major alkaloid from saprophytic cultures of *Aspergillus fumigatus* Fres. alongside fumigaclavine B and festuclavine[79]. The total alkaloid yield was 150 mg/liter of culture. Fumigaclavine A and B have not previously been observed as products of the ergot fungus.

Fumiclavine A crystallizes from aqueous methanol colorless needles, m.p. 84–85°. Fumigaclavine A gives an intense blue color with the reagent of *Allport* and *Cocking*[106].

Hydrochloride crystallizes from ethanol in prisms, m.p. 304–305° (dec.).

$$[\alpha]_D^{20} = -56.7° \ (c = 1.5 \ in \ methanol)$$

Fumigaclavine B
$$C_{16}H_{20}ON_2 \ (m.w. \ 256.3)$$

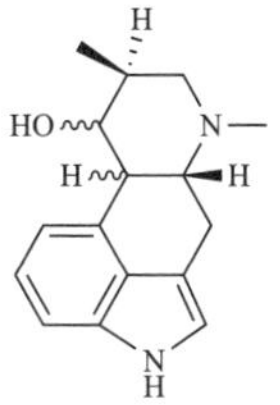

Fumigaclavine B was obtained as a minor alkaloid in saprophytic cultures of *Aspergillus fumigatus* Fres.[79]. It was also detected by paper chromatography in cultures of *Rhizopus arrhizus* Fischer. It is formed from fumigaclavine A by hydrolytic cleavage of the acetyl group. From aqueous alcohol needles, m.p. 244–245°, at 260° the melt becomes solid and then melts again at 265–267°.

$$[\alpha]_{5461}^{20} = -\ 6.3° \ (c = 1.2 \ in \ methanol)$$
$$[\alpha]_{5461}^{20} = -\ 113° \ (c = 0.6 \ in \ pyridine)$$

UV spectrum: maxima at 225 mμ (log ε_{max} = 4.49), 275 mμ (log ε_{max} = 3.79), 282 mμ (log ε_{max} = 3.82), 293 mμ (log ε_{max} = 3.84).

Methiodide from methanol/ether prisms, m.p. 310–311° (dec.).

Chanoclavine

$C_{16}H_{20}ON_2$ (m.w. 256.3)

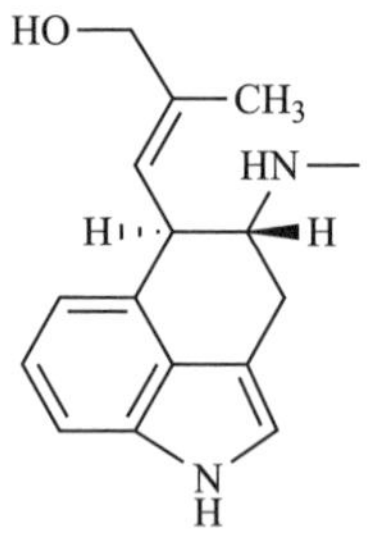

Chanoclavine (for origin of name see Section IV 2 h) was first found in saprophytic cultures of the ergot fungus of *Pennisetum typhoideum* Rieb.[18]. It occurs in small amounts in ergot of various provenances and in the saprophytic cultures prepared from it[19, 35]. The alkaloid X[72] isolated by *M. Abe* et al. form the ergot of various grasses (*Elymus, Phragmites, Phalaris, Agropyrum*), formulated as dihydroelymoclavine and later named secaclavine[74], seems to have been, according to the divergent physicochemical data, not yet uniform chanoclavine.

Chanoclavine is also one of the ergot alkaloids found in the seeds of morning glory (*Rivea corymbosa* [L.] Hall. f., *Ipomoea tricolor* Cav.)[78]. Chanoclavine crystallizes from methanol or acetone in solid solvent-free prisms and polyhedra, m.p. 220–222° (dec.). The alkaloid dissolves at boiling heat in 25 parts methanol, 140 parts acetone, 170 parts ethyl acetate, or 350 parts chloroform.

$$[\alpha]_D^{20} = -240°; [\alpha]_{5461}^{20} = -294° \ (c = 1.0 \ in \ pyridine)$$
$$pK^* = 8.2 \ (in \ 80\% \ aq. \ methyl \ cellosolve)$$

UV spectrum: maxima at 225 mµ (log ε_{max} = 4.44), 284 mµ (log ε_{max} = 3.82), 293 mµ (log ε_{max} = 3.76)

IR spectrum: See Section V, 3b, Figure 15z, 2.

Color reactions: Violet-blue coloration with Keller or van Urk reagent, same as dihydrolysergic acid and its derivatives.

Hydrogen oxalate from water or methanol needles, m.p. 195–197° (dec.), $[\alpha]_D^{20} = -152° \ (c = 0.5 \ in \ 50\% \ alcohol)$.

N-acetylchanoclavine from methanol massive prisms, m.p. 226–228° (dec.), $[\alpha]_D^{20} = -80° \ (c = 0.5 \ in \ pyridine)$.

O,N-diacetylchanoclavine from benzene/petroleum ether, needles stubbornly retaining 1 mol crystal benzene, m.p. 174–175° (dec.), $[\alpha]_D^{20} = -80° \ (c = 0.5 \ in \ pyridine)$.

2. Structural elucidation of the alkaloids of the clavine group

The structural elucidation of the clavine alkaloids consists primarily in their conversion into derivatives of the lysergic acid and dihydrolysergic acid series, which resulted in a structural and steric linkage with these compounds whose constitution was already ensured.

The experimentally performed transformations from the clavine series to the lysergic acid series and the connections within the individual alkaloids of the clavine group can be seen in Formula Scheme 21. In the formula diagrams, the correct configurations are already shown, as they were established according to the absolute configuration of lysergic acid determined since then by physical and chemical methods (see Section III, 2e).

a)   Connection of elymoclavine with dihydro-*d*-lysergic acid

and with agroclavine

The catalytic hydrogenation of elymoclavine resulted in a mixture of dihydro-*d*-lysergol-(I) and dihydro-*d*-isolysergol-(I)[180]. Thus, elymoclavine has the same ring system as dihydro-*d*-lysergic acid and the same configuration at C5 and C10, i.e., rings C and D are *trans*-linked, and at C5 the spatial arrangement is the same as in *d*-lysergic acid.

Reduction of elymoclavine with sodium in butanol gave a mixture of agroclavine, lysergine, pyroclavine, festuclavine, and costaclavine, besides dihydro-*d*-lysergol-(I) and dihydro-*d*-isolysergol-(I)[174, 181]. The location of the isolated double bond in agroclavine, for which the 7–8 or the 8–9 position was in question, was derived, among other things, from the fact that the agroclavine showed practically the same pK value as its dihydro derivative[182].

Heating agroclavine in sodium butyrate solution gave exclusively lysergine, while the same treatment for elymoclavine resulted in a mixture of lysergol and lysergine. Lysergol, like agroclavine or lysergine, yielded festuclavine as the only product in the catalytic reduction[181].

The constitution of elymoclavine was clearly established on the basis of the conversions discussed above. Not so clear was the derivation of the configuration of agroclavine, which was based on the reduction of elymoclavine to agroclavine with sodium in butanol. Since this energetic reaction also yielded, in addition to agroclavine, reaction products that were formed with the removal of the asymmetry center at C10, the possibility of a *cis*-linkage of rings C and D in agroclavine and in the other clavine alkaloids derived from it was not excluded.

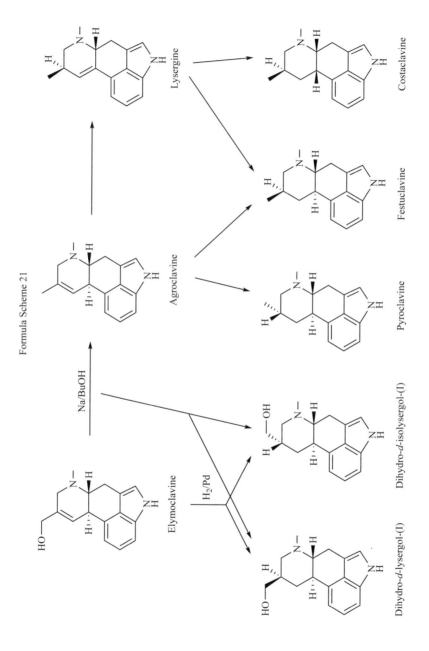

Formula Scheme 21

Lysergine

Costaclavine

Agroclavine

Festuclavine

Pyroclavine

Na/BuOH

Elymoclavine

H₂/Pd

Dihydro-*d*-isolysergol-(I)

Dihydro-*d*-lysergol-(I)

Formula Scheme 22

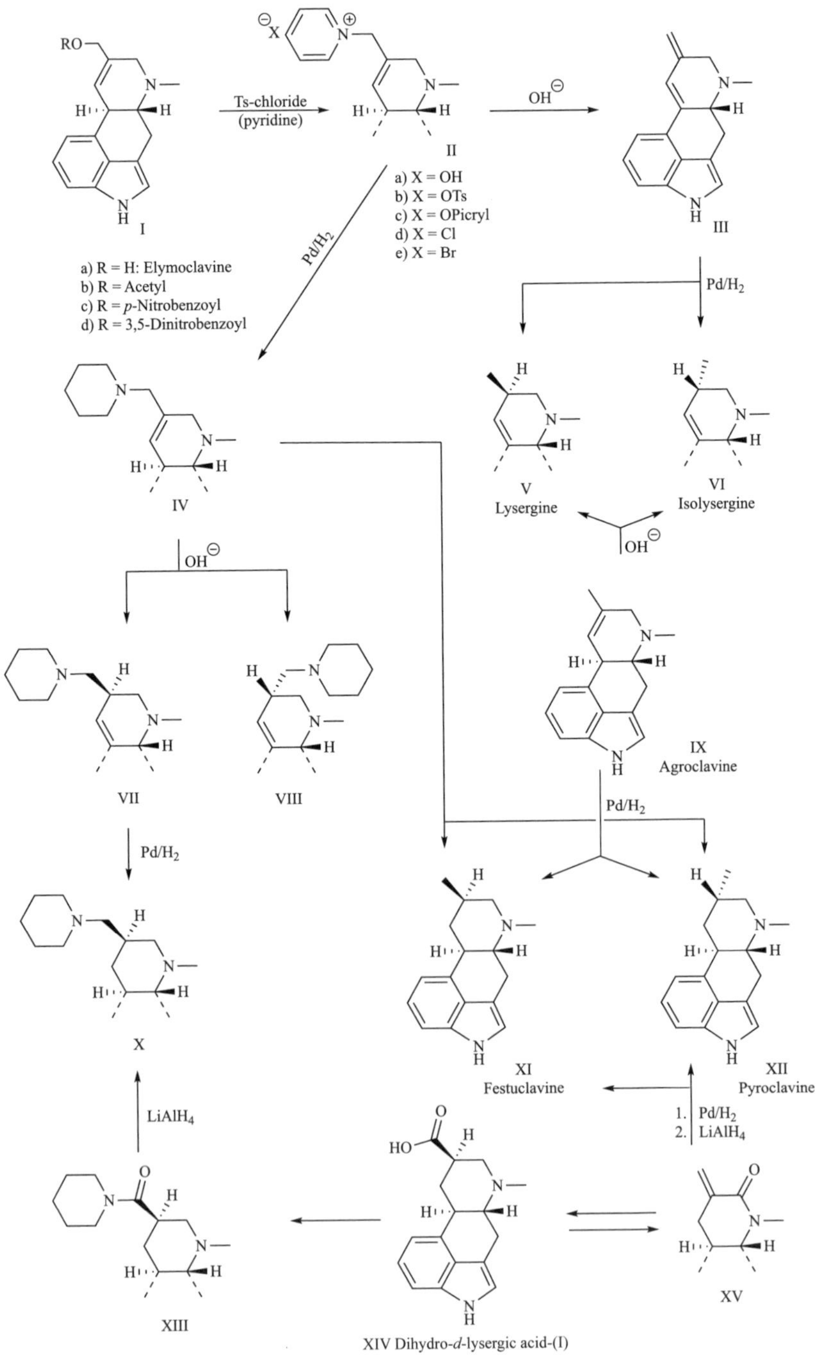

I
a) R = H: Elymoclavine
b) R = Acetyl
c) R = p-Nitrobenzoyl
d) R = 3,5-Dinitrobenzoyl

II
a) X = OH
b) X = OTs
c) X = OPicryl
d) X = Cl
e) X = Br

Ts-chloride (pyridine)

OH⊖

Pd/H₂

III

Pd/H₂

IV

V
Lysergine

VI
Isolysergine

OH⊖

OH⊖

VII

VIII

IX
Agroclavine

Pd/H₂

Pd/H₂

X

XI
Festuclavine

XII
Pyroclavine

1. Pd/H₂
2. LiAlH₄

LiAlH₄

XIII

XIV Dihydro-d-lysergic acid-(I)

XV

The correctness of the above derivations is confirmed by the unambiguous conversions recorded in Formula Scheme 22[175].

b) Connection of agroclavine with elymoclavine

1. Reaction sequence: $Ia \rightarrow II \rightarrow IV \rightarrow XI + XII \leftarrow IX$

Elymoclavine Ia could be converted via steps II and IV, while maintaining the conformation at C5 and C10, into the 6,8-dimethylergoline XI (festuclavine) and XII (pyroclavine), which are diastereomeric at C8. On the other hand, XI and XII were also obtained from agroclavine (IX) by catalytic reduction without changing the centers of asymmetry at C5 and C10. Thus, elymoclavine and agroclavine are sterically unambiguously linked, and it is proved that elymoclavine and agroclavine match in their spatial structure.

The first intermediate II (X = OTs) in the conversion of elymoclavine to XI and XII was obtained in an attempt to esterify elymoclavine in pyridine solution with tosyl chloride. It is known from the literature that sulfonic acid esters of allyl alcohols readily react with pyridine to form quaternary salts. In the present case, the reactivity of the intermediately formed tosyl ester was so great that it was not possible to capture it, and the readily water-soluble pyridinium tosylate was obtained directly.

The reductive conversion of II to XI and XII can be divided into two steps by partial hydrogenation, where IV (d-6-methyl-8-piperidinomethyl-$\Delta^{8,9}$-ergolene) can be captured. Analogous to the known reductive cleavage of benzylamines, an allylamine cleavage occurs during the transition from IV to XI and XII.

2. Reaction sequence: $Ia \rightarrow IIb \rightarrow III \rightarrow V + V$

A further connection of elymoclavine with agroclavine was achieved by converting the two alkaloids into the 6,8-dimethylergolenes V (lysergine) and VI (isolysergine). Boiling II with dilute alkali gave III (lysergene) in good yield, which was converted to a mixture of V and VI by partial hydrogenation.

c) Connection of agroclavine with dihydro-*d*-lysergic acid-(I)

Reaction sequence: *IX → XI + XII ← XV ← XIV*

The stereoisomeric 6,8-dimethylergolines XI and XII, which can be obtained from agroclavine as described under b), could also be prepared from dihydro-*d*-lysergic acid-(I) (XIV) via the previously described lactam[92] (XV) by a two-step reduction, first with Pd-C/hydrogen in alcohol and then with $LiAlH_4$, which also linked agroclavine to the dihydrolysergic acid series.

d) Connection of elymoclavine with dihydro-*d*-lysergic acid-(I)

Reaction sequence: *Ia → IIb → IV → VII → X ← XIII ← XIV*

As a common reaction product, which could be prepared from both elymoclavine and dihydro-*d*-lysergic acid-(I) by unique reactions, *d*-6-methyl-8-piperidinomethyl-ergoline-(I) (X) was chosen for the linkage. Compound IV, already mentioned in b), could be alkali-isomerized by shifting the isolated double bond into a mixture of VII and VIII. VII (*d*-6-methyl-8-piperidinomethyl-$\Delta^{9,10}$-ergolene) gave X (*d*-6-methyl-8-piperidinomethyl-ergoline-(I)) upon catalytic reduction with Pt in glacial acetic acid, which was also obtained by $LiAlH_4$ reduction of dihydro-*d*-lysergic acid piperidide (XIII).

e) Structural elucidation of setoclavine and isosetoclavine; connection with agroclavine

The oxidation of agroclavine with potassium dichromate in diluted sulfuric acid[180] gives a mixture of setoclavine and isosetoclavine[18] (for formulae see Section IV, 1). The two alkaloids have the same UV spectrum as the lysergic acid and isolysergic acid, which shows that the isolated double bond of agroclavine has migrated to the 9–10 position conjugated to the indole system. The oxygen that has entered is present in a non-acetylatable, i.e., tertiary hydroxyl group.

In the $\Delta^{9, 10}$-ergolene substituted in position 8, a tertiary hydroxyl can be placed only at C8. The detection of a C-methyl group in setoclavine and isosetoclavine, comparison of pK values and chromatographic behavior, allowed setoclavine and isosetoclavine to be formulated as lysergine and isolysergine, respectively, hydroxylated in the 8-position.

f)   Structural elucidation of penniclavine and isopenniclavine; connection with elymoclavine

The ring structure of penniclavine and isopenniclavine (formulae, see Section IV, 1) resulted from the conversion of elymoclavine by oxidation with potassium dichromate in dilute sulfuric acid into these two alkaloids[18, 180]. Their UV spectra agree with those of lysergic acid, revealing the $\Delta^{9, 10}$-ergolene skeleton. Further insight into their structure was provided by periodate oxidation. Both isomers released formaldehyde with this reagent, demonstrating the grouping -$CHOHCH_2OH$. However, a glycol grouping with a primary hydroxyl could be attached to the ergolene skeleton postulated for penniclavine and isopenniclavine only in the 8-position. The isomers with the weaker basicity and stronger retention to the Alox-column were based on the regularities identified in this series assigned the lysergic acid configuration, and the diastereomer on C8 was assigned the iso configuration. This gave penniclavine the constitution d-8-hydroxylysergol, and isopenniclavine the constitution d-8-hydroxyisolysergol.

g)   Structural elucidation of fumigaclavine A and B

Characteristic of fumigaclavine A is an acetyl group, which is easily cleaved by alkali, forming fumigaclavine B. Fumigaclavine B yields fumigaclavine A again upon acetylation. The light deacetylation and the IR bands at 1,241 cm^{-1} and 1,725 cm^{-1} indicated an ester grouping. From the UV spectrum, it could be concluded that there was no double bond conjugated to the indole system. Heating of fumigaclavine B with sodium hydroxide afforded anhydro-fumigaclavine B, which was identical to lysergine. From these data and conversions, the following structures could be derived for fumigaclavine A and B.

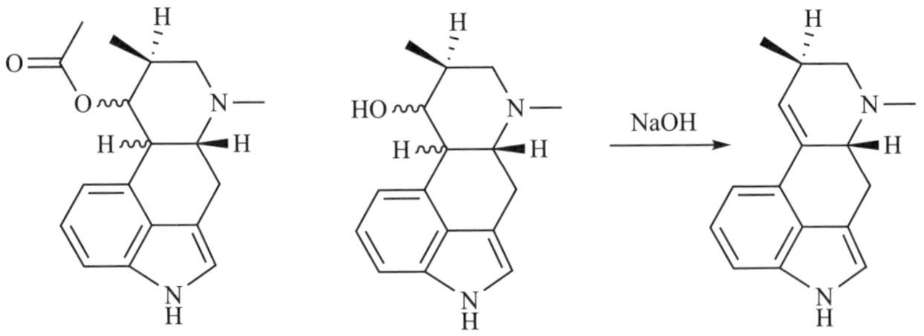

| Fumigaclavine A | Fumigaclavine B | Lysergine |

In these formulas, the configurations at C9 and C10 are open. The negative optical rotation of fumigaclavine A and B in pyridine suggests a *trans*-linkage of the rings C and D.

h) Structural elucidation of chanoclavine

This alkaloid was so named, because the ring D of the ergoline system is split into an open chain in it. While all other ergot alkaloids known to date are based on the tetracyclic ergolene or ergoline skeleton, chanoclavine is derived from the tricyclic 1,3,4,5-tetrahydrobenz[c,d]indole.

Unlike the other alkaloids of the clavine group, chanoclavine can be acetylated very easily with acetic anhydride in pyridine. In this process, two acetyl groups enter the molecules. Diacetylchanoclavine is no longer basic, which shows that the basic nitrogen atom of the chanoclavine is acetylatable, that is, primary or secondary. The second acetyl group is located at an easily esterifiable, that is primary, hydroxyl group. In the IR spectrum of diacetylchanoclavine, the distinct band of the acid amide group is visible at 1,630 cm^{-1}, and the characteristic absorption of the acetoxy group at 1,740 cm^{-1}. This can be readily alkali hydrolyzed to give the N-monoacetylchanoclavine.

Chanoclavine has the same number of C atoms as lysergic acid and the other *Pennisetum* alkaloids, which argued against its formulation as a 6-nor-compound with ring D intact. As another possible variant with acetylatable nitrogen, only one formula remained, in which the ring D between the nitrogen and the C atom 7 is open. The ring D is also opened at this position when the lysergic acid or the dihydrolysergic acid is heated with acetic anhydride[92]. The empirical formula of the chanoclavine requires a double bond in the carbon side chain when ring D is open. As the UV spectrum showed, it must not be conjugated with the indole system. Thus, only the 7–8 or the 8–9 position could be considered. Since chanoclavine has a $C \cdot CH_3$ group, only the $\Delta^{9, \, 10}$-position remained for the isolated double bond. All these findings could be expressed by the following structural formula of chanoclavine.

The configuration at the two centers of asymmetry C5 and C10 resulted from the connection of chanoclavine with festuclavine. The latter compound was formed in low yield in an attempt to selectively hydrogenate chanoclavine catalytically.

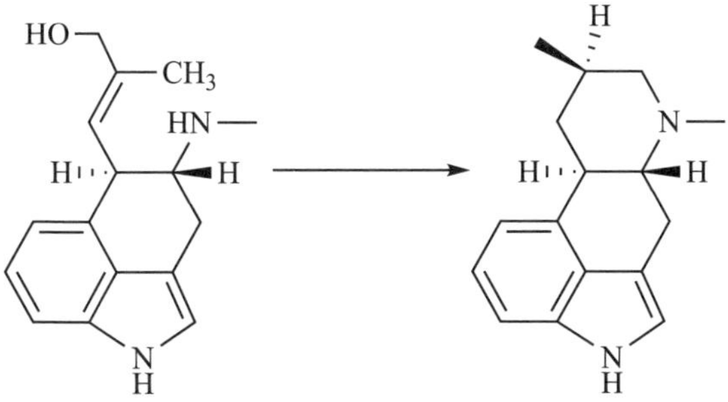

Chanoclavine   Festuclavine

i)   Notes on the structure of molliclavine

Since this alkaloid has been isolated in one single laboratory and only in very small quantity[73], only a provisional structural formula can be assigned to it on the basis of the sparse measurements and reactions that have been carried out with it. The UV spectrum agrees with that of agroclavine elymoclavine. While the alkaloid itself gives a green color with Keller or van Urk reagent, its dihydro derivative gives a violet-blue color. The empirical formula $C_{16}H_{18}O_2N_2$ and biogenetic considerations, together with the stated findings, led to the hypothetical formulation of molliclavine as 9-hydroxyelymoclavine.

V. Analytics of ergot alkaloids

The extraction of ergot alkaloids from ergot or from saprophytic cultures, the purification of the alkaloid fraction, the separation into the individual alkaloids, and their purification can be carried out by the methods commonly used in alkaloid chemistry. When working with ergot alkaloids, however, these methods must be adapted to the easy decomposability of these substances, especially their sensitivity to oxidative influences and exposure to light and, in the case of some of the alkaloids, their easy isomerizability.

The qualitative and quantitative determination methods of the total alkaloids and the individual ergot alkaloids are based almost entirely on their indole nature by exploiting the characteristic color reactions and absorption spectra of this class of substances. For special purposes, titration, especially in anhydrous medium, is also used for quantitative measurements. Recent reviews of methods for the determination of ergot alkaloids in the drug and in pharmaceutical preparations have been published by *G. E. Foster*[183] and by *J. Gyenes* and *J. Bayer*[184].

1. Description of some typical extraction procedures

Finely ground ergot, defatted with petroleum ether, is moistened with dilute ammonia and then extracted with peroxide-free ether in a continuous extraction apparatus. From the ethereal extract, the total alkaloids are transferred to the aqueous phase by shaking out with 1% tartaric acid solution. In the tartaric acid solution, the total amount of ergot alkaloids can be quantitatively determined colorimetrically by one of the methods described below[185].

A gentler extraction method consists of mixing the defatted ergot powder with sodium bicarbonate and shaking out the moistened mass with ether containing 5% ethanol[186].

Degreasing of the ergot is unnecessary if the drug is extracted with tartaric acid–containing 50% aqueous methanol or acetone. After evaporation of the organic solvent in vacuo and alkalinization, the alkaloids can be taken up into the organic phase with ether, chloroform, or ethyl acetate.

Another extraction method, which is used less for analytical purposes than for the preparative recovery of alkaloids, has already been described as a process for the industrial production of ergotamine (see Section C, III/1).

2. Color reactions of ergot alkaloids

a) Keller's color reaction

The first characteristic color test for ergot alkaloids was published in 1896 by *C. C. Keller*[187]. It consists of dissolving a trace of the alkaloid in glacial acetic acid containing a small amount of $FeCl_3$ and underlaying the solution with conc. sulfuric acid (pipetting the acid below the $FeCl_3$ layer). A violet-blue ring then forms at the boundary layer of the two phases. When shaken, the solutions turn blue to green. However, the intensity and nuance of the coloration is very dependent on the purity of the glacial acetic acid used and the amount of $FeCl_3$ added, so this color reaction could not be used for quantitative measurements for the time being.

However, by suitable standardization, a method could be developed which allows the alkaloids of the ergotamine group to be qualitatively distinguished from those of the ergotoxine group in a simple manner. Even the individual alkaloids within the ergotoxine group can be distinguished from one another.

Also for the qualitative differentiation of individual Rauwolfia alkaloids and Rauwolfia alkaloid groups, the standardized Keller's color reaction proved quite useful[188]. *H. P. Rieder* and *M. Böhmer*[189, 190] have shown that the glyoxylic acid contained in glacial acetic acid plays a decisive role in this color reaction.

When the conditions described in Table 14 are met, the ergot alkaloids give the colorations indicated there[301].

After hydrogenation, the alkaloids of the ergotoxine group can no longer be distinguished from the other ergot alkaloids by the above color reaction. All derivatives of dihydrolysergic acid, without distinction, give a consistent blue color, as opposed to the coloration of the lysergic acid derivatives shifted by a nuance to violet.

b) Color reaction according to van Urk-Smith

All ergot alkaloids give a blue coloration in sulfuric acid solution with p-dimethylaminobenzaldehyde. This color reaction was first described by *H. W. van Urk*[191] and later improved by *M. I. Smith*[192] for quantitative measurements. It has been adapted in a form standardized by *N. L. Allport* and *T. T. Cocking*[106], in which a specific amount of ferric chloride is added to accelerate the color development, in most pharmacopoeias for the quantitative colorimetric determination of ergot alkaloids.

Table 14

Differentiating color reactions of ergot alkaloids.

Dissolve 0.2-0.3 mg alkaloid (mol wt. approx. 600) or 0.1-0.15 mg alkaloid (mol wt. approx. 300) in 1.0 ml glacial acetic acid containing 0.5‰ Fe^{III} in the form of $FeCl_3$ and 0.1% glyoxylic acid, carefully underclay with 1.0 ml conc. sulfuric acid (nitrate-free) and shake after 15 seconds.

Alkaloid	Color reaction	
Lysergic acid alkaloids	Color change	Final color after 2 hours
Ergotamine-Ergotaminine	blue	blue
Ergosine-Ergosinine	blue	blue
Ergocristine	from blue in 10-15 seconds to olive green	olive green
Ergocryptine	from blue in 1 minute to dark green; after 2 minutes green; after 30 minutes olive green	olive green
Ergocornine	from blue in 2 minutes to greenish blue; after 10 minutes blue-green	green

Ergocristinine	from blue in 15-20 seconds to olive green	olive green
Ergocryptinine	from blue in 1 minute to green-blue; after 3 minutes blue-green	olive green
Ergocorninine	from blue in 5 minutes to greenish blue	green
Ergobasine (ergometrine) and all other mono derivatives of lysergic lysergic acid and isolysergic acid	blue	blue
Setoclavine	green	olive green
Isosetoclavine	olive green	yellow green
Penniclavine	green	green
Isopenniclavine	olive green	yellow green
Other alkaloids of the clavine group	blue	blue

According to this procedure, 0.125 g of p-dimethylaminobenzaldehyde is dissolved in a cooled mixture of 65 ml of sulfuric acid and 35 ml of water to which 0.1 ml of a 5% $FeCl_3$ solution has been added. For the measurement, 1.0 ml of the aqueous tartaric acid solution to be tested is mixed with 2.0 ml of this reagent and the resulting blue coloration is compared with a standard solution in a colorimeter at a wavelength of 550 nm. According to US Pharmacopoeia XVI and British Pharmacopoeia 1958, ergonovine bimaleinate serves as the standard preparation. Since only the lysergic acid part is responsible for the coloration, the values found for other alkaloids must be converted according to the mol. wt.

The pharmacologically less active isolysergic acid derivatives give qualitatively and quantitatively the same coloration as the active lysergic acid alkaloids. The van Urk-Smith's color reaction thus cannot be used to distinguish the active from the inactive alkaloids.

Several authors have dealt with the specificity of this color reaction and its quantitative evaluation[193–196, 377]. It has also been elaborated for the determination of the alkaloid content in individual sclerotia[197, 198]. A comparison of van Urk's color reaction with the Keller's color reaction and with the staining with conc. sulfuric acid was made by S. *Yamatodani*[199].

On the basis of the broad agreement of the absorption spectra of the blue colorations produced in the van Urk-Smith color reaction of ergot alkaloids on the one hand, and on the other hand in the reaction of ß-indolylacetic acid with *p*-dimethylaminobenzaldehyde, the following limiting forms were rendered probable for the blue color salt[200].

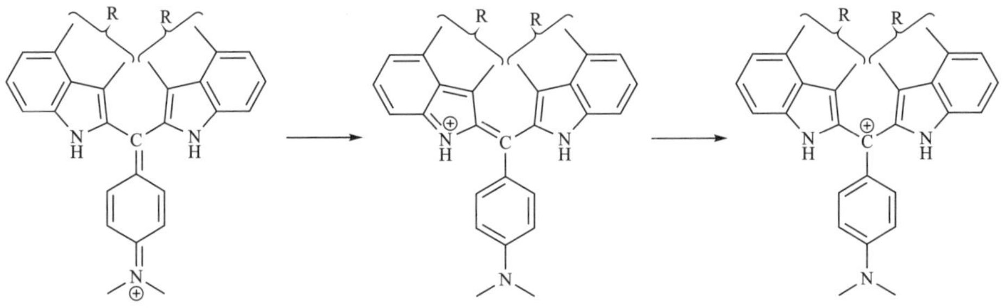

3. Spectral analytical data

a) UV spectra

All ergot alkaloids derived from lysergic acid, as well as the alkaloids of the clavine group with an ergolene-(9) system, give the same characteristic UV absorption spectrum, which is characterized by a flat maximum at $\lambda = 316-318$ mµ and a minimum at $\lambda = 268$ mµ (see Figure 14). This spectrum is due to the indole system and its conjugated carbon double bond in the 9–10 position and is therefore also present in all derivatives of isolysergic acid and isoergolene-(9).

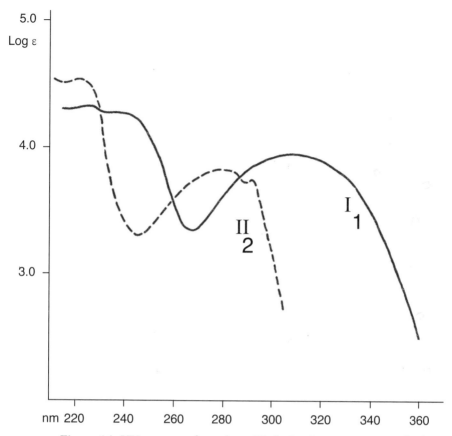

Figure 14. UV spectra of ergolene-(9) derivatives = curve I and of ergoline derivatives = curve II. Solvent: methanol

The measurement of the absorbance at the absorption maximum allows the quantitative determination of individual alkaloids or of the total amount of alkaloids present in a mixture, but does not allow a differentiation between the individual alkaloids or the lysergic acid and isolysergic acid forms any more than the spectrophotometric evaluation of the van Urk-Smith color reaction[184, 201, 202].

When the double bond in the 9–10 position of the ergolene system is saturated, the absorption maximum shifts by about 30 nm, into the short-wavelength region. All dihydrolysergic acid derivatives, as well as the lumi-compounds (see Figure 19) and the ergolene derivatives with an isolated carbon double bond in the 8–9 position show the characteristic absorption spectrum of the indole system with the maximum at 282–284 nm and a secondary maximum at 292 nm and a minimum at 245 nm (Figure 14). With the aid of the UV spectrum, contamination of lysergic acid alkaloids with dihydro derivatives and vice versa can be conveniently determined.

b) IR spectra

The IR spectra of ergot alkaloids are, like those of any other compounds, excellently suited for the identification of single constituents and for purity testing of preparations. They have not yet been used for quantitative determinations and are probably less suitable than the UV spectra or the colorimetric methods.

The IR spectra of the more important natural ergot alkaloids are compiled in Figure 15[203]. The spectra were recorded with a Perkin-Elmer instrument Mod. 237. The spectral half-width was 5 cm^{-1} in the region of 3000 cm^{-1} and 2 cm^{-1} below 1700 cm^{-1}.

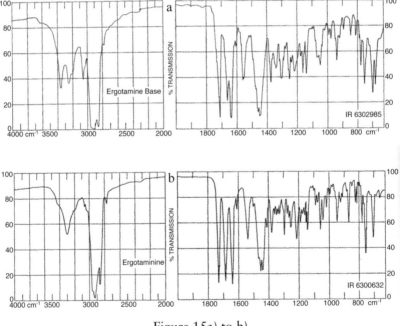

Figure 15a) to b)
IR spectra of ergot alkaloids (in Nujol)

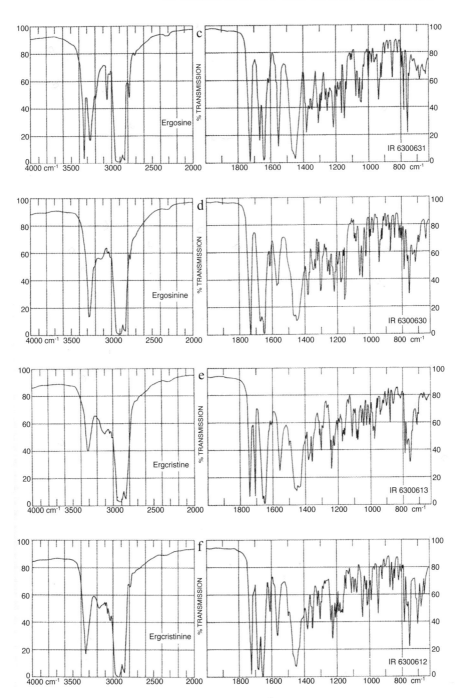

Figure 15c) to f)
IR spectra of ergot alkaloids (in Nujol)

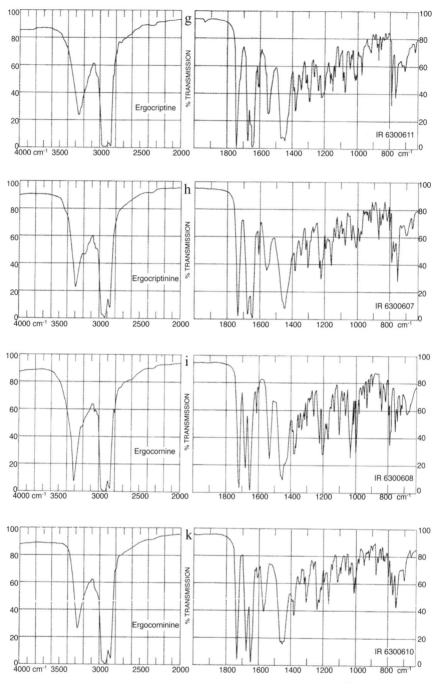

Figure 15g) to k)
IR spectra of the ergot alkaloids (in Nujol)

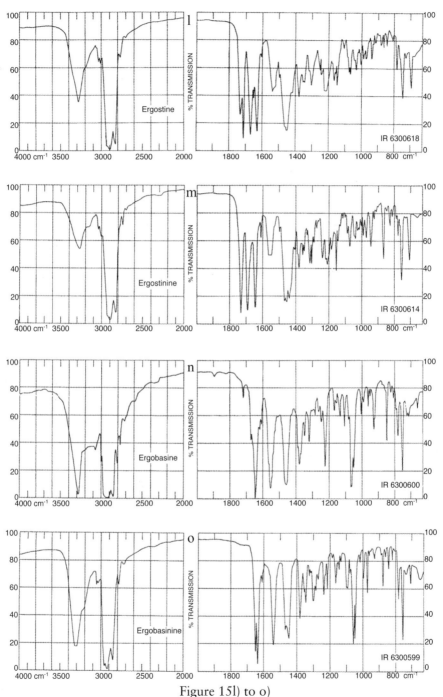

Figure 15l) to o)
IR spectra of ergot alkaloids (in Nujol)

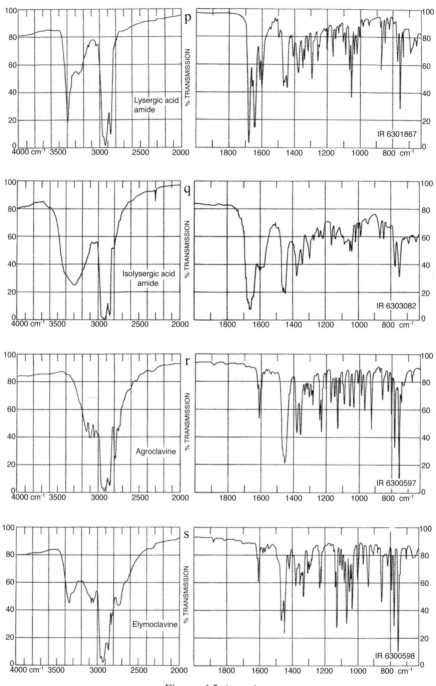

Figure 15p) to s)
IR spectra of ergot alkaloids (in Nujol)

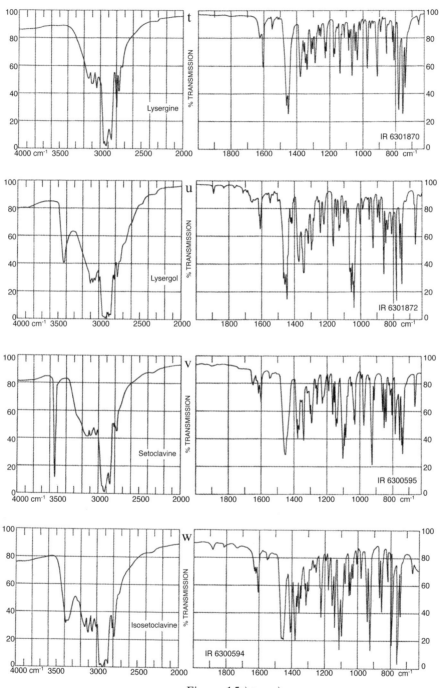

Figure 15t) to w)
IR spectra of the ergot alkaloids (in Nujol)

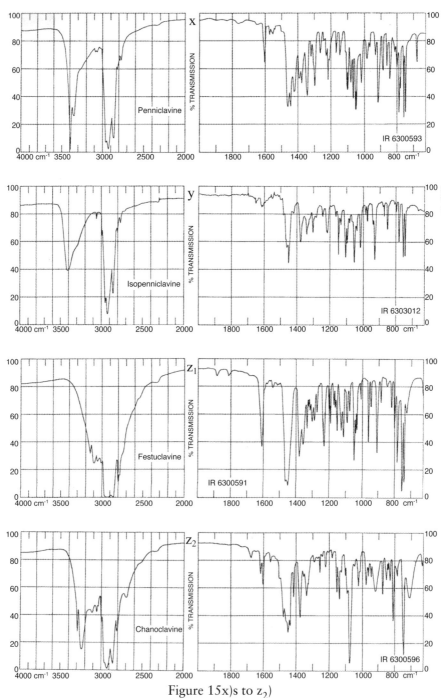

Figure 15x)s to z$_2$)
IR spectra of the ergot alkaloids (in Nujol)

c) Fluorescence spectra

All ergot alkaloids that have the double bond conjugated to the indole system in the 9–10 position fluoresce with blue color in UV light. When this double bond is hydrogenated, the fluorescent color disappears in visible light[120], as it also does when it is saturated with water[204]. The development of a new apparatus, the spectro-photofluorometer by *R. L. Bowman* et al.[205] (manufacturer: American Instrument Co., Washington), allowed the extension of fluorescence analysis to compounds that fluoresce in the UV range. Therefore, with the Aminco-Bowman apparatus or equivalent instruments, ergot alkaloids and derivatives lacking a double bond in the 9–10 position that do not fluoresce in visible light can also be determined spectrophotofluorometrically.

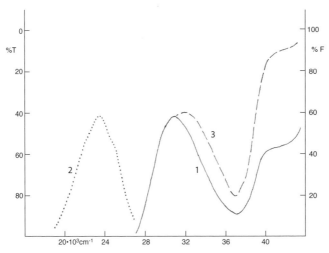

Figure 16. Emission spectrum (1), fluorescence spectrum (2) (after excitation at 325 nm), UV absorption spectrum (3) of lysergic acid amide (ergine). Solvent: ethanol

In principle, there are two ways of using the fluorescence of a compound, either in visible light or in the UV range, for analytical determination:

a)   The fluorescence generated by excitation with light of a specific wavelength can be measured. Advantageously, monochromatic light of the wavelength at which the substance in question shows an absorption maximum is used for excitation. This method is suitable for the quantitative determination of a particular compound.

b)   For qualitative characterization, but also for quantitative measurements, both the excitation and emission spectra are used, which are produced when the wavelength of the excitation light changes continuously. Since the conversion to fluorescence radiation is greatest at points of maximum absorption, the emission spectra largely correspond to the absorption spectra. It is therefore possible, with very small amounts of substance in the order of µg, with which UV absorption spectrum cannot be recorded anymore, to obtain characteristic emission spectra corresponding to the absorption curve.

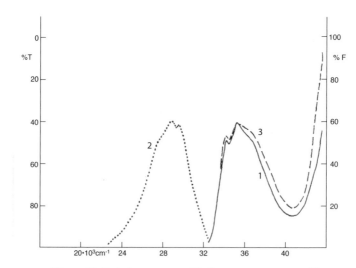

Figure 17. Emission spectrum (1), fluorescence spectrum (2)
(after excitation at 285 nm), UV absorption spectrum
(3) of 9,10-dihydrolysergic acid amide. Solvent: ethanol

Figures 16–18 show the fluorescence spectrum, UV absorption spectrum, and emission spectrum (excitation spectrum) of three typical ergot alkaloids, lysergic acid amide (ergine) (= $\Delta^{9\text{-}10}$-ergolene derivative), dihydrolysergic acid amide (= ergoline derivative), and agroclavine (= $\Delta^{8\text{-}9}$-ergolene derivative)[299].

The extraordinary sensitivity of spectrophotofluorometric methods makes them particularly suitable for the determination of ergot alkaloids and their derivatives in biological material[298]. Thus, with the help of fluorescence measurement, the distribution in animal tissue and the biochemical transformation of lysergic acid diethylamide (Lysergsäurediäthylamid, LSD) was determined[208]. Under excitation at 325 nm and measurement of fluorescence light at 445 nm, concentrations as low as 0.001 µg per ml of lysergic acid diethylamide could be determined.

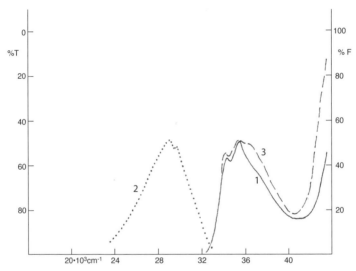

Figure 18. Emission spectrum (1), fluorescence spectrum (2) (after excitation at 285 nm), UV absorption spectrum (3) of agroclavine. Solvent: ethanol

4. Separation methods

a) Fractional extraction

After the discovery of ergobasine as a water-soluble, purely oxytocically active component of ergot, the analytical evaluation of the drug could no longer be limited to the determination of the water-insoluble alkaloids of the ergotamine and ergotoxine groups.

The first method for dividing the total ergot alkaloids into a water-soluble and a water-insoluble portion was described by C. H. *Hampshire* and C. R. *Page*[185, 209]. It is based on the fact that the water-soluble bases can be extracted from an ethereal solution of the total alkaloids with weakly ammoniacal water.

This method of separating the two alkaloid groups after the extraction procedure was later improved in other laboratories by adjusting the aqueous phase to a particular pH[210, 211, 294] and also elaborated for the analysis of individual sclerotia[213–215].

Fractional extraction in the form of *Craig*'s countercurrent distribution has also been applied to the partitioning of ergot alkaloid mixtures[216, 217].

b) Column chromatography

The mixtures of ergot alkaloids can be separated chromatographically into individual alkaloids and alkaloid groups on columns of suitable adsorbents, especially on Alox columns. With chloroform or methylene chloride, to which increasing amounts of methanol or alcohol are added, the alkaloids of the lysergic acid series can be eluted in the following order:

a) Dextrorotatory alkaloids of the ergotoxine and ergotamine groups and ergotoxine complex (with pure $CHCl_3$ or CH_2Cl_2)

b) Ergotamine and ergosine (after addition of 0.5% methanol or ethanol)

c) Ergobasine (after addition of 5% methanol or ethanol).

The separation into these three groups is achieved by using about 50 times the amount of aluminum oxide (*Brockmann*) relative to the alkaloid mixture to be separated. Rapid work protected from light is necessary, since the alkaloids on Alox are decomposed particularly easily[202, 218].

Group separation is also possible with silica gel columns[378] or with cellulose powder columns. In the most recent publication by R. *Voigt* on this process[287], the earlier experiments are also discussed.

Column chromatography is well suited for the preparative separation of alkaloid mixtures into groups. On the contrary, clean separation of the ergotoxine complex is not successful with this method. It can, however, be achieved by paper chromatography.

c) Paper chromatography

The paper chromatographic method was initially used for the separation of ergobasine and ergobasinine[186]. Subsequently, numerous systems for the paper chromatographic separation of ergot alkaloid mixtures were described, cited in the 1954 work of *A. Stoll* and *A. Rüegger*, who succeeded in neatly separating the ergotoxine–ergotinine complex and the ergotamine–ergotaminine group into the 6 and 4 individual alkaloids, respectively[290]. They used paper strips impregnated with phthalic acid dimethyl ester (Whatman No. 1) as the stationary phase and developed with a formamide-water mixture of different concentration and pH. Ergobasine and ergobasinine and the mixture of hydrogenated ergot alkaloids also separated well by this method, as can be seen from Table 15 of the Rf values.

The separation of the non-hydrogenated alkaloids can be observed by their intense blue-violet fluorescence in UV light. In the final chromatogram, the bases with their blue-violet color reaction according to van Urk are visualized by soaking the dried sheets with a solution of p-dimethylaminobenzaldehyde in cyclohexane and, after evaporation of the solvent, placing them in an atmosphere of HCl gas. Less than 1 µg alkaloid is still detectable in this way.

For the analysis of complex mixtures of ergot alkaloids, it is advantageous to separate them first into groups on the Alox or cellulose column and these then further resolve by paper chromatography.

Paper chromatography has still been used in multiple variants for the separation and determination of ergot alkaloids[379], e.g., by the two-dimensional method[292], further with quantitative evaluation[206, 207, 302], or specifically for the determination of the alkaloids of the clavine group[296, 380], the hydrogenated alkaloids[381] and the lumi ergot alkaloids[382].

Table 15

Rf values of ergot alkaloids

	Composition of the aqueous phase and its pH adjusted with formic acid	Rf values
Ergocornine		0.73
Ergocryptine		0.63
Ergocristine	Formamide-water 4:6 (v/v) pH: 4.0	0.51
Ergocorninine		0.35
Ergocryptinine		0.25
Ergocristinine		0.17
Ergotamine		0.37
Ergosine	Formamide-water 4:6 (v/v) pH: 5.2	0.55
Ergotaminine		0.16
Ergosinine		0.27
Ergobasine	Formamide-water 1:9 (v/v) pH: 5.0	0.52
Ergobasinine		0.63
Dihydroergocornine		0.57
Dihydroergocryptine	Formamide-water 1:4 (v/v) pH: 4.0	0.44
Dihydroergocristine		0.31

It provided valuable services in the determination and identification of isomerization products in pharmaceutical preparations[103] and for the determination of ergot alkaloids in biological material, e.g., ergometrine in viscera[286].

d) Thin-layer chromatography

Thin-layer chromatography, which was further developed and became generally known through the investigations of E. Stahl[391], has already been used in many variants for the separation of ergot alkaloids. Using aluminum oxide and silica gel as solid phase material and chloroform with the addition of methanol as an eluent, the ergot alkaloids occurring in small quantities in the seeds of winch species (*Ipomoea tricolor* Cav. and *Rivea corymbosa* [L.] Hall. f.) could be separated and determined[78]. This method has also been advantageously used for the study of the alkaloids of paspalum ergot[19] and for the separation of the clavine alkaloids[291]. As on the paper chromatograms, the substance spots on the thin-layer plate can be identified by their fluorescence in UV light and by their violet-blue color when developed with p-dimethylaminobenzalde-hyde and HCl gas.

Because of the simplicity and low time requirements of this method, thin-layer chromatography is quite suitable for serial studies and for testing ergot alkaloid preparations for uniformity.

VI. Biogenesis of ergot alkaloids

The tryptamine unit, belonging to the large class of indole alkaloids, is incorporated, as in almost all plant bases, in the ring skeleton of the lysergic acid and of the clavine alkaloids.

In all hypotheses about the biogenesis of the lysergic-acid part of the ergot alkaloids, tryptophan was therefore considered as the initial stage. Subsequently, it could be shown experimentally that this assumption was correct.

Radioactive ergot alkaloids were produced after the injection of D,L-tryptophan-[ß-^{14}C] into the internodes of rye plants following previous inoculation with *Claviceps purpurea*. The lysergic and isolysergic acids obtained from this by degradation were radioactive[212]. In a field experiment, an incorporation rate of 0.158% of the applied radiolabeled tryptophan was found. The incorporation of tryptophan could also be detected in saprophytic cultures. After addition of D,L-tryptophan-[ß-^{14}C] to an elymoclavine and agroclavine producing fungal strain, incorporation rates of 10% to 39% were found, with addition of pyridoxal phosphate improving the yield. Experiments with tryptophan-[^{14}COOH] yielded practically inactive alkaloids, from which it can be concluded that the carboxyl group of the tryptophan does not enter the ergoline skeleton during biosynthesis[289]. Also, in incorporation experiments with a *Claviceps purpurea* strain that produced lysergic acid alkaloids in vitro, very high incorporation rates of D,L-tryptophan-[ß-^{14}C] were observed. The high specific activities of the isolated alkaloids indicated that also

D-tryptophan was used[288]. This was then proven by direct incorporation of tritium-labeled D-tryptophan, 14% of which was used by saprophytic cultures for the synthesis of elymoclavine[295]. In submerged cultures of *Claviceps paspali* Stevens and Hall, in which mainly lysergic and isolysergic acid amide is formed, D,L-tryptophan-[ß-14C] could be incorporated into the alkaloid fraction at 54%[367].

Since tryptophan can be built up from indole and serine by *Claviceps purpurea*[300], indole-[2-14C] could also be incorporated into the alkaloids in saprophytic cultures of ergot fungus. The incorporation rate of 1.12, resp. 1.39%, into ergosine could be increased to 1.88% by simultaneous addition of serine. These experiments also showed that tryptophan is used as a whole by the fungus for alkaloid synthesis and is not predegraded to smaller fragments. Such degradation would result in the loss of C2 atom of the indole system, which, however, as shown by the incorporation of indole-[2-14C], is retained during biosynthesis. The finding that tryptamine-[ß-14C] is not used by the fungus makes decarboxylation of the tryptophan prior to incorporation into the alkaloid molecule unlikely. The incorporation rate of 1.35% for L-methionine-[methyl-14C] in festuclavine and pyroclavine seems to indicate that the N-methyl group of ergot alkaloids is taken over from methionine via a transmethylation reaction[297].

Tryptophan specifically deuterated in the 5- or 6-position was used in saprophytic cultures of the *Pennisetum* ergot fungus to build up the clavine alkaloids without loss of deuterium, whereas deuterium was lost in the 4-position. These experiments showed that the hypotheses according to which 5-hydroxytryptophan is passed through as an intermediate during biosynthesis[219, 293] could not be correct[220].

Through the studies discussed above, it had been proven that tryptophan is a building block in the biogenesis of the ergoline system. Various hypotheses have been put forward about the nature of the residue required for further assembly into the carbon-nitrogen skeleton of lysergic acid.

According to *E. E. van Tamelen*[219], dihydro-nornicotinic acid or a derivative thereof should be considered for this purpose.

J. Harley-Mason[293] postulated acetonedicarboxylic acid (2-oxomalo-nic acid) and formaldehyde as possible precursors. *N. L. Wendler*[221] assumed that lysergic acid is formed by condensation of tryptophan with citric acid. According to *R. Robinson*[222], the ring closure is supposed to come about by succinic acid and a 1-C body. According to *A. Feld-stein*[223], tryptophan should cyclize and then condense with α-keto acid to form lysergic acid. *K. Mothes* et al.[212] assumed that an isoprenoid 5-C component is involved in the formation of ergoline. While for the first five hypotheses no experimental documentation was provided, the last mentioned assumption could be proven by experiments, and it could be shown that the ergoline system in the fungus is actually synthesized according to the following structure scheme.

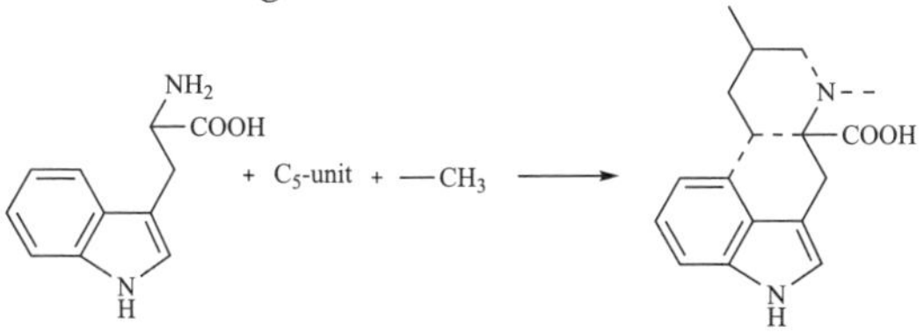

Labeled mevalonic acid, which had been identified as a precursor of active isoprene, was used for the incorporation experiments. *D. Gröger* et al.[224] achieved an incorporation rate of 16.4% by adding mevalonic acid-[2-^{14}C] in phosphate buffer together with L-tryptophan to submerged cultured, filtered mycelium of a mainly elymoclavine-producing *Claviceps* strain. Using tritiated mevalonic acid-[2T] or mevalonic acid-[4T], they found specific incorporation rates of 22.2% and 23.5%, or 18.0% and 9.2%. In a strain producing ergotamine, ergosine, and ergometrine in saprophytic culture, *E. H. Taylor* and *E. Ramstad*[225] also observed a high incorporation rate of radioactive carbon when mevalonic acid-[2-^{14}C] was used, and like-wise in a clavine alkaloid producing fungal strain[383]. These findings allowed the assumption that mevalonic acid, or an active isoprene residue resulting from it, is actually involved in the biosynthesis of the ergoline system.

Biogenetic studies with subsequent localization of the labeled C-atom in the active alkaloids allowed a more precise statement about the nature of the incorporation of mevalonic acid, resp. the resulting isoprenoid, into the ergoline system. *A. J. Birch* et al.[226, 227] conducted their studies on elymo-clavine and agroclavine obtained by adding mevalonic acid-[2-^{14}C] to the nutrient solution of a *Claviceps* strain of *Pennisetum typhoideum*. The two alkaloids were degraded according to the following Formula Scheme 23.

The acetone obtained from agroclavine, which consists of C atoms 7, 8, and 17, contained all the activity, and the methylamine corresponding to C17 had 78% of the activity. It follows that the *C2 of mevalonic acid enters the molecule as the C17 of agroclavine. The findings for ely-moclavine led to the same result. 86% of the activity was found in the formaldehyde, which was formed from the C17 of the alkaloid.

Corresponding results were obtained by *Baxter* et al.[228, 384] on radio-labeled lysergic acid from ergosine obtained by feeding mevalonic acid-[2-^{14}C]. The CO_2 corresponding to C17 formed during *Schmidt* degradation contained practically all the activity, while the C7 isolated via the lysergic acid lactam as formaldehyde, resp. CO_2, was almost inactive. The meva-lonic acid is thus incorporated into the lysergic acid molecule according to Formula Scheme 24 on the next page.

Formula Scheme 23

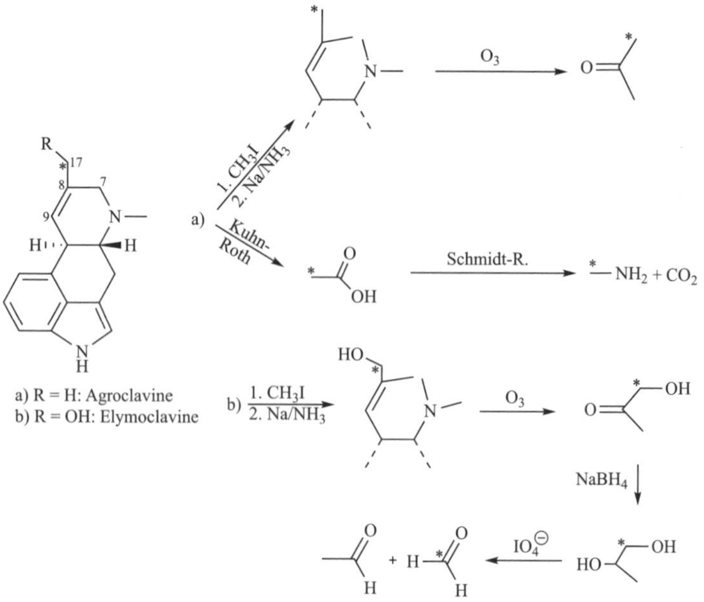

a) R = H: Agroclavine
b) R = OH: Elymoclavine

Addition of mevalonic acid-[1-^{14}C] to a pyroclavine and festuclavine producing fungal strain resulted in inactive alkaloids, which is in agreement with the above scheme showed that the carboxyl of mevalonic acid is not incorporated. Decrease of the incorporation rates of mevalonic acid-[2-^{14}C] by addition of dimethyl allyl pyrophosphate or isopentenyl pyrophosphate reinforced the view that mevalonic acid enters the molecule via one of these activated isoprene residues. This could also be directly deduced from the experiment.

Formula Scheme 24

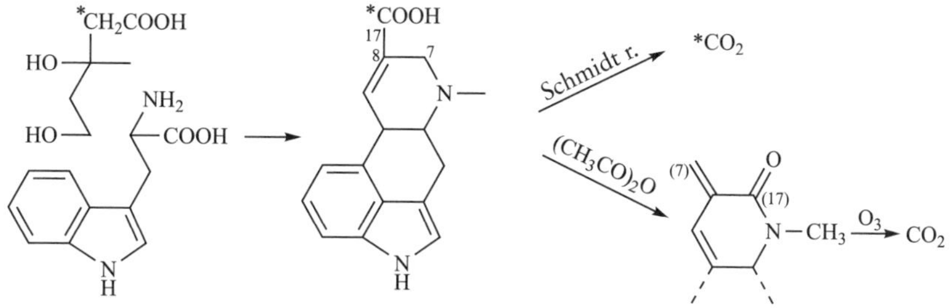

Diffusely deuterated isopentenyl pyrophosphate was utilized by a *Claviceps* strain in saprophytic culture for the synthesis of clavine alkaloids[220].

After the experiments discussed above had identified mevalonic acid, resp. an active isoprene residue arising from it, as a biogenetic building block of the ergoline system, it remained to be clarified how this residue combines with the tryptamine moiety. *F. Weygand* et al.[229] found that in surface cultures of a *Claviceps purpurea* strain producing primarily elymoclavine, 1-(ß-indolyl)-2-amino-5-methyl-hexene-(4) labeled with tritium in the indole moiety was incorporated to 20–30% into this alkaloid. These findings could not be confirmed later. A corresponding precursor, with the dimethyl allyl residue in the 4-position of the tryptophan, ^{14}C-labeled in the ß-position of the alanine side chain, showed, although in a different fungal strain, an approximately 10 times smaller incorporation rate[230].

Finally, *R. M. Baxter* et al.[358] showed that when D,L-mevalonic acid-[2-^{14}C -5-T] was incorporated into festuclavine and pyroclavine, one atom of hydrogen was lost, whereas when D,L-mevalonic acid-[2-^{14}C-2-T] was used for the biosynthesis of agroclavine, the degree of oxidation at C2 of mevalonic acid was preserved.

It seems that the linkage of the active isoprene residue (dimethyl allyl pyrophosphate) to position 4 of the tryptophan is achieved by electrophilic attack from C5 of the isoprene unit with loss of the pyrophosphate ion, and then by oxidation of the CH_2 group (= C5 of the isoprene unit), a secondary hydroxyl group is then formed, which again becomes a potential electrophilic center by pyrophosphorylation. The electrophilic carbon atom formed by cleavage of the pyrophosphate ester could then establish the bond with the α-C atom of the tryptophan side chain.

The next step in elucidating the biogenesis of ergot alkaloids was to determine the biogenetic relationship between the individual ergot alkaloids.

In the clavine series, the addition of ^{14}C-labeled alkaloids to saprophytic cultures showed that agroclavine converts to elymoclavine, penniclavine, and isopenniclavine, whereas only penniclavine and isopenniclavine were formed from labeled elymoclavine. Labeled penniclavine and isopenniclavine themselves did not cause any formation of elymoclavine or agroclavine[231, 363]. Thus, at least for this fungal strain, it was proven that the biogenesis pathway consists of a progressive hydroxylation. Experiments on the course of alkaloid formation by the *Claviceps* strain of *Pennisetum typhoideum* by *A. Brack* et al. also point in this direction, showing that the content of penniclavine and setoclavine increases as the culture period progresses with a simultaneous decrease in elymoclavine and agroclavine[232], and the finding that lysergol and lysergene labeled with tritium are not incorporated into agroclavine[358].

Furthermore, *K. Mothes* et al. found an incorporation of elymoclavine-[T] into ergotamine in experiments on sclerotia and an incorporation of elymoclavine-[T] and elymoclavine-[^{14}C] into *d*-lysergic acid amide[364] in submerged cultures of *Claviceps paspali*, which makes it seem possible that at least part of the lysergic acid alkaloids is formed in this way via clavine alkaloids during natural biosynthesis. Also from the comprehensive work of R. Voigt[386–389], in which the alkaloid spectrum of parasitic and saprophytic cultures was studied under different conditions and in different stages of development, it is essentially evident that the biosynthesis of ergot alkaloids seems to lead from lower to higher oxidation levels.

Further elucidation of biogenesis and its finer mechanisms is the main topic of ongoing research in the field of ergot alkaloids[390].

VII. Derivatives of the ergot alkaloids

The valuable pharmacological properties of the natural parent alkaloids prompted their chemical modification. Numerous derivatives were prepared and tested pharmacologically and to some extent also clinically in order to elucidate the relationship between chemical structure and physiological activity in this substance class, and in the hope of finding substances with new effects useful in therapy. This has indeed been successful, as will be shown below. This line of research will be pursued further, because the possibilities for variation offered by the ergot alkaloid molecule in chemical and pharmacological terms are far from exhausted.

1. Acid amide-like derivatives of lysergic acid and dihydrolysergic acid

The azide procedure used in the first partial synthesis of ergobasine and its stereoisomeric forms (cf. Section C, III/4) was also used to prepare homologs and other closely related derivatives of this alkaloid[157, 233, 234]. Tables 16 and 17 summarize such derivatives of ergobasine.

From the series of compounds listed in Tables 16 and 17, d-lysergic acid (+)-butanolamide-(2), the next-highest homolog of ergobasine in the amino alcohol side chain, has become medically important. This substance is widely used under the brand name "Methergine"® in obstetrics as a uterotonic and hemostatic agent. As with ergobasine (cf. Section C, III/4), all eight theoretically possible stereoisomeric forms of lysergic acid butanolamide have also been prepared[235].

Table 16
Homologues and derivatives of ergobasine

Designation	Empirical formula (Mol. wt.)	M.p. (dec.)	$[\alpha]_D^{20}$
d-Lysergic acid ethanolamide (Nor-ergobasine)	$C_{18}H_{21}O_2N_3$ (311.4)	95°	$-10°$ (in pyridine)
d-Isolysergic acid ethanolamide (Nor-ergobasinine)		206°	$+448°$ (in pyridine)
d-Lysergic acid (+)-butanolamide-(2) (Methylergobasine)	$C_{20}H_{25}O_2N_3$ (339.4)	172°	$-45°$ (in pyridine)
d-Isolysergic acid (+)-butanolamide-(2) (Methylergobasinine)		194°	$+386°$ (in CHCl$_3$)
d-Lysergic acid L-3-methyl-butanolamide-(2) (Dimethyl-ergobasine)	$C_{21}H_{27}O_2N_3$ (353.4)	115°	$-59°$ (in pyridine)
d-Isolysergic acid L-3-methyl-butanolamide-(2) (Dimethylergobasinine)		188°	$+463°$ (in pyridine)
d-Lysergic acid L-4-methyl-pentanolamide-(2) (Isopropyl-ergobasine)	$C_{22}H_{29}O_2N_3$ (367.5)	130°	$-38°$ (in pyridine)
d-Isolysergic acid L-4-methyl-pentanolamide-(2) (Isopropylergobasinine)		160°	$+330°$ (in CHCl$_3$)
d-Lysergic acid 1,3-dihydroxy-propanamide-(2) (Hydroxy-ergobasine)	$C_{19}H_{23}O_3N_3$ (331.4)	125°	$+55°$ (in water)
d-Isolysergic acid 1,3-dihydroxy-propanamide-(2) (Hydroxyergobasinine)		231°	$+445°$ (in pyridine)
d-Lysergic acid L-N-benzyl-propanolamide-(2) (N-benzylergobasine)	$C_{26}H_{29}O_2N_3$ (415.5)	230°	$-17°$ (in pyridine)
d-Lysergic acid l-ephedride (N-methylphenyl-ergobasine)	$C_{26}H_{29}O_2N_3$ (415.5)	258°	$-21°$ (in pyridine)

Table 17
Stereoisomers of phenylergobasine and cyclopentylergobasine

Compound	M.p. (dec.)	$[\alpha]_D^{20}$ (acetone)
Phenylergobasines: $C_{25}H_{27}O_2N_3$ (401.5)		
d-Lysergic acid-d-nor-ephedride	230°	$+14°$
l-Lysergic acid-l-nor-ephedride	230°	$-16°$
d-Isolysergic acid-d-nor-ephedride	128°	$+267°$
l-Isolysergic acid-l-nor-ephedride	128°	$-267°$
d-Lysergic acid-l-nor-ephedride	130°	$-17°$
d-Isolysergic acid-l-nor-ephedride	130°	$+296°$
d-Lysergic acid-d-nor-ψ-ephedride	128°	$+27°$
d-Isolysergic acid-d-nor-ψ-ephedride	-	$+370°$
Cyclopentylergobasines: $C_{24}H_{31}O_2N_3$ (393.5)		$[\alpha]_D^{20}$ (pyridine)
d-Lysergic acid-(+)-3-cyclopentyl-l-hydroxy-2-propylamide	120°	$+14°$
d-Isolysergic acid-(+)-3-cyclopentyl-l-hydroxy-2-propylamide	153°	$+450°$
d-Lysergic acid-(-)-3-cyclopentyl-l-hydroxy-2-propylamide	192°	$-4°$
d-Isolysergic acid-(-)-3-cyclopentyl-l-hydroxy-2-propylamide	203°	$+356°$

154

Furthermore, a large number of amide-like derivatives of lysergic acid and dihydrolysergic acid have been synthesized in which the amino alcohol residue of ergobasine has been replaced by other simple or more complicated primary and secondary amines, and which are thus related to the natural ergot alkaloids only in that they have the lysergic acid nucleus in common with them.

Table 18 lists a number of monosubstituted, and Table 19 disubstituted, amides of lysergic acid and isolysergic acid with simple amine components. Table 20 contains the data for some amides of the isomeric dihydrolysergic acids[121, 157, 233].

Table 18

Monosubstituted amides of lysergic acid and isolysergic acid

$C_{15}H_{15}N_2 \cdot COR$

R =	Substituent		Empirical formula (Mol. wt.)	M.p. (dec.)	$[\alpha]_D^{20}$
—NH₂	*l*-Lysergic acid *l*-Isolysergic acid	amide	$C_{18}H_{21}O_2N_3$ (267.3)	132-134°	~0° −489°
—N⟨H	*d*-Lysergic acid *d*-Isolysergic acid	methylamide	$C_{17}H_{19}ON_3$ (281.3)	– 183°	+26° +445°
—N⟨H	*d*-Lysergic acid *d*-Isolysergic acid	ethylamide	$C_{18}H_{21}ON_3$ (295.4)	153-155° 188°	−3° +476°
—N⟨H	*d*-Lysergic acid *d*-Isolysergic acid	propylamide	$C_{19}H_{23}ON_3$ (309.4)	– 118°	−4° +472°
HN—	*d*-Lysergic acid *d*-Isolysergic acid	isopropylamide	$C_{19}H_{23}ON_3$ (309.4)	200-210° –	−4° +432°
—N⟨H	*d*-Lysergic acid *d*-Isolysergic acid	*n*-butylamide	$C_{20}H_{25}ON_3$ (323.4)	110-120° 140-143°	−10° +430°
—N	*d*-Lysergic acid *d*-Isolysergic acid	*n*-amylamide	$C_{21}H_{27}ON_3$ (335.4)	190° 153°	−12° +409°
—NH	*d*-Lysergic acid *d*-Isolysergic acid	allylamide	$C_{19}H_{27}ON_3$ (307.4)	– 156°	+1° +439°
—N⟨H	*d*-Lysergic acid (+)-β-phenyl-isopropylamide		$C_{19}H_{23}O_3N_3$ (385.5)	–	−13°
—NH	*d*-Lysergic acid (+)-β-(trimethoxyphenyl)ethylamide (*d*-Lysergic acid mescalide)		$C_{27}H_{31}O_4N_3$	185°	−8°

The optical antipodes of the *l*-lysergic acid amide and *l*-isolysergic acid amide covered in Table 18, the *d*-lysergic acid amide and the *d*-isolysergic acid amide, have already been described as naturally occurring ergot alkaloids under the names ergine and erginine (cf. Section C, III/1).

Table 19

Disubstituted amides of lysergic acid and isolysergic acid, $C_{15}H_{15}N_2 \cdot COR$

R =	Substitutent	Empirical formula (Mol. wt.)	M.p. (dec.)	$[\alpha]_D^{20}$
—N<	*d*-Lysergic acid ⎱ dimethylamide *d*-Isolysergic acid ⎰	$C_{18}H_{21}ON_3$ (295.4)	— 146°	− 29° + 199°
—N⊏	*d*-Lysergic acid *d*-Isolysergic acid ⎱ diethylamide *l*-Lysergic acid ⎰ *l*-Isolysergic acid	$C_{20}H_{25}ON_3$ (323.4)	80-85° 182° 80-85° 182°	+ 17° + 219° − 17° −219°
—N⟨⟩	*d*-Lysergic acid di-*n*-propylamide	$C_{22}H_{29}ON_3$ (351.5)	195°	+ 18°
—N⟨	*d*-Lysergic acid ⎱ diisopropylamide *d*-Isolysergic acid ⎰	$C_{22}H_{29}ON$ (351.5)	246° 260°	+ 35° + 178°
—N⟨⟩	*d*-Lysergic acid di-*n*-butylamide	$C_{24}H_{33}ON_3$ (379.5)	97°	+ 15°
—N⟨⟩	*d*-Lysergic acid ⎱ (+)-methyl-(β-phenyl-isopropyl)-*d*-Isolysergic acid ⎰ amide	$C_{26}H_{29}ON$ (399.5)	186° 212°	+ 21° + 174°
—N⟨⟩	*d*-Lysergic acid ⎱ piperidide *d*-Isolysergic acid ⎰	$C_{21}H_{25}ON$ (335.4)	— 140°	+ 18° + 190°
—N⟨⟩O	*d*-Lysergic acid ⎱ morpholide *d*-Isolysergic acid ⎰	$C_{20}H_{23}O_2N$ (337.4)	116° 188°	+ 26° + 203°
—N⟨⟩	*d*-Lysergic acid ⎱ pyrrolidide *d*-Isolysergic acid ⎰	$C_{20}H_{23}ON$ (321.4)	181° 203°	+ 25° + 193°

In the series of disubstituted amides of lysergic acid, *d*-lysergic acid diethylamide is characterized by extraordinarily high and specific hallucinogenic and psycholytic activity. This compound has become known under the abbreviated name LSD-25, brand name "Delysid"®, in experimental psychiatry and is now gaining increasing importance in psychotherapy as a medicinal aid. In Section D, the pharmacology and clinical application of the partially synthetic, acid amide-like derivatives of lysergic acid are discussed in more detail.

Table 20

Amides of the isomeric dihydrolysergic acids, $C_{15}H_{17}N_2 \cdot COR$

R =	Substituent	Empirical formula (Mol. wt.)	M.p. (dec.)	$[\alpha]_D^{20}$ (pyridine)
—NH₂	Dihydrolysergic acid amide	C₁₆H₁₉ON₃ (269.2)	267°	−131°
	Dihydroisolysergic acid-(I) amide		275°	0°
	Dihydroisolysergic acid-(II) amide		307°	+17°
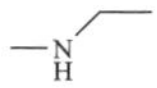 —N< (H)	Dihydrolysergic acid ethylamide	C₁₈H₂₃ON₃ (297.4)	251°	−136°
	Dihydroisolysergic acid-(I) ethylamide		239°	+22°
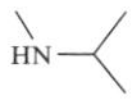 HN—<	Dihydrolysergic acid isopropylamide	C₁₉H₂₅ON₃ (311.4)	255°	−140°
 —N N—	Dihydrolysergic acid-[N-methylpiperidyl-(4)] amide	C₂₂H₃₀ON₄ (366.5)	276°	−130°
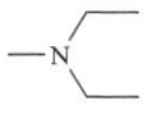 —N<	Dihydrolysergic acid diethylamide	C₂₂H₂₇ON₃ (325.4)	130-135°	−114°
	Dihydroisolysergic acid-(I) diethylamide		240-243°	−68°
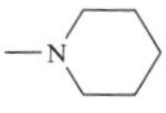 —N⟨ ⟩	Dihydrolysergic acid piperidide	C₂₁H₂₇ON₃ (337.4)	197°	−122°
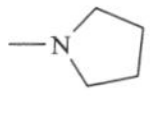 —N⟨ ⟩	Dihydrolysergic acid pyrrolidide	C₂₀H₂₅ON₃ (323.4)	214°	−119°
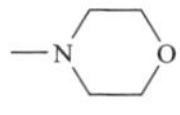 —N⟨ O⟩	Dihydrolysergic acid morpholidide	C₂₀H₂₅O₂N₃ (339.4)	156-158°	−112°

Tables 21 and 22 contain homologous series of cycloalkylamides[236] and (cyclopentylalkyl)amides[237] of lysergic acid and isolysergic acid.

Some of the lysergic acid derivatives listed in Tables 21 and 22 exhibit remarkable pharmacodynamic properties. The cyclobutyl and cyclopentyl amides of d-lysergic acid show high oxytocic activity on the rabbit uterus in vivo[238].

The anti-serotonin effect on isolated organs is also particularly pronounced with these two compounds.

In an effort to obtain active ingredients with the pharmacological qualities of the peptide alkaloids of the ergotamine and ergotoxine series, a large number of peptide-like derivatives of lysergic acid and dihydrolysergic acid have been prepared, in which they are linked to amino acids or di- and tripeptides[239, 240]. Tables 23–26 summarize the most important chemical and physical data of such derivatives.

Table 21

Cycloalkylamides of d-lysergic acid and d-isolysergic acid

Substituent		Empirical formula (Mol. wt.)	M.p. (dec.)	$[\alpha]_D^{20}$ (pyridine)
d-Lysergic acid	cyclopropylamide	$C_{19}H_{21}ON_3$	199-201°	−11.6°
d-Isolysergic acid		(307.4)	175-176°	+470°
d-Lysergic acid	cyclobutylamide	$C_{20}H_{23}ON_3$	121-122°	−16.4°
d-Isolysergic acid		(321.4)	202-204°	+452°
d-Lysergic acid	cyclopentylamide	$C_{20}H_{25}ON_3$	121-122°	−27.9°
d-Isolysergic acid		(335.4)	233-235°	+463°
d-Lysergic acid	cyclohexylamide	$C_{22}H_{27}ON_3$	131-133°	−33.3°
d-Isolysergic acid		(349.5)	204-205°	+449°
d-Lysergic acid	cycyloheptylamide	$C_{23}H_{29}ON_3$	125-128°	−41.4°
d-Isolysergic acid		(363.5)	185-186°	+439°

Table 22

(ω-Cyclopentylalkyl)amides of d-lysergic acid and d-isolysergic acid

$$C_{15}H_{15}N_2-\overset{\overset{O}{\|}}{C}-\underset{\underset{H}{|}}{N}-(CH_2)_n-\bigcirc$$

Substituent			Empirical formula (Mol. wt.)	M.p. (dec.)	$[\alpha]_D^{20}$ (pyridine)
n=		Row			
1	Lysergic acid		$C_{22}H_{27}ON_3$	95-98°	−18.5°
	Isolysergic acid		(349.5)	185-186°	+414°
2	Lysergic acid		$C_{23}H_{29}ON_3$	115-116°	−11°
	Isolysergic acid		(363.5)	150°	+392°
3	Lysergic acid		$C_{24}H_{30}ON_3$	165-166°	−12°
	Isolysergic acid		(377.5)	amorphous	+322°
4	Lysergic acid		$C_{25}H_{33}ON_3$	150-151°	−7°
	Isolysergic acid		(391.5)	amorphous	+289°

Table 23
Peptides associated with lysergic acid

Amino acid or peptide component	Empirical formula (Mol. wt.)	$[\alpha]_D^{20}$	Typical crystallization and m.p. (dec.)
Glycine amide	$C_{18}H_{20}O_2N_4$ (324.4)	- 6° (pyridine)	From acetone when diluted with benzene in round crystal aggregates, m.p. 152°
Glycine diethylamide	$C_{22}H_{28}O_2N_4$ (380.5)	- 8° (pyridine)	Base amorphous. Acid maleate from methanol in needles, m.p. 193°
L-alanine methyl ester	$C_{20}H_{23}O_3N_3$ (353.4)	- 63° (CHCl₃)	Base amorphous. Bioxalate from methanol in needles, m.p. 188-194°
α-Aminobutyric acid ethyl ester	$C_{22}H_{27}O_3N_3$ (381.5)	- 45° (CHCl₃)	Base amorphous. Bioxalate from acetone in roundish aggregates, m.p. 159-162°
L-phenylalanine amide	$C_{25}H_{26}O_2N_4$ (414.5)	- 60° (CHCl₃)	Hydrochloride from alcohol in rosettes, m.p. over 250° unsharp
L-Leucine ethyl ester	$C_{24}H_{31}O_3N_3$ (409.5)	- 71° (CHCl₃)	Base amorphous. Bioxalate from acetone in roundish aggregates, m.p. 185-188°
L-Tryptophane methyl ester	$C_{28}H_{28}O_3N_4$ (468.6)	- 23° (CHCl₃)	Base amrphous. Hydrochloride needles, m.p. 190-193°
Glycyl-L-leucine methyl ester	$C_{25}H_{32}O_4N_4$ (452.5)	- 68° (CHCl₃)	Base amorphous. Bioxalate from acetone fine crystals, m.p. 180-185°

The hope of finding among these substances those with sympatholytic properties was not fulfilled. Apparently, for the specific pharmacological effects of the alkaloids of the ergotamine and ergotoxine group and their dihydro derivatives to occur, the cyclol structure of the peptide moiety is

Table 24
Peptides associated with isolysergic Acid

Amino acid or peptide component	Empirical formula (Mol. wt.)	$[\alpha]_D^{20}$	Typical crystallization and m.p. (dec.)
Glycine amide	$C_{18}H_{20}O_2N_4$ (324.4)	+ 237° (pyridine)	From chloroform needle tufts, m.p. 120-130°
L-Alanine methyl ester	$C_{20}H_{23}O_3N_3$ (353.4)	+ 310° (CHCl₃)	Base amorphous. Bioxalate from acetone spherical aggregates, m.p. 164-169°
α-Aminobutyric acid ethyl ester	$C_{22}H_{27}O_3N_3$ (381.5)	+ 296° (CHCl₃)	Base amorphous. Bioxalate from acetone circular aggregates, m.p. 159-162°
L-Leucine ethyl ester	$C_{24}H_{31}O_3N_3$ (409.5)	+ 309° (CHCl₃)	Base amorphous. Bioxalate m.p. 160-162°
L-Leucine diethyl amide	$C_{26}H_{36}O_2N_4$ (436.6)	+ 222° (CHCl₃)	Base amorphous. Bioxalate m.p. 156-159°
L-Phenylalanine methyl ester	$C_{26}H_{27}O_3N_3$ (429.5)	+ 278° (CHCl₃)	From benzene solid pointed prisms, m.p. 140 160°
L-Tryptophane methyl ester	$C_{28}H_{28}O_3N_4$ (468.6)	+ 197° (CHCl₃)	Did not crystallize
L-Histidine methyl ester	$C_{23}H_{25}O_3N_5$ (419.6)	+ 224° (CHCl₃)	Did not crystallize
Glycyl-L-leucine methyl ester	$C_{25}H_{32}O_4N_4$ (452.5)	+ 266° (CHCl₃)	Base amorphous. Bioxalate m.p. 140-144°

necessary. In contrast, some of these substances showed high uterotonic activity in animal studies. For example, dihydrolysergyl glycinamide is six times more effective in situ on the rabbit uterus than d-lysergic acid (+)-butanolamide-(2) (Methergine®). The oxytocic activity of dihydrolysergyl glycinamide could be further enhanced by the introduction of 1 or 2 bromine atoms, but these substances proved to be only weakly effective or even ineffective in human clinical trials[241].

All compounds included in Tables 16–26 have been prepared by the azide method. In recent years, other methods for the partial synthesis of ergobasine and other amide-like derivatives of lysergic acid have become known.

Table 25
Peptides associated with dihydrolysergic acid $C_{15}H_{17}N_2 \cdot COR$

Amino acid or peptide component	Empirical formula (Mol. wt.)	$[\alpha]_D^{20}$	Typical crystallization and m.p. (dec.)
L-Alanine	$C_{19}H_{23}O_3N_3$ (341.4)	- 65° (5% aq. NH₃)	From aqueous NH₃, needles, m.p. 356°
L-Proline	$C_{21}H_{25}O_3N_3$ (367.4)	- 122° (5% aq. NH₃)	From aqueous NH₃, needles, m.p. 321°
Glycyl glycine	$C_{20}H_{24}O_4N_4$ (384.4)	- 50° (5% aq. NH₃)	From aqueous NH₃, needles, m.p. 239°
Glycine amide	$C_{18}H_{22}O_2N_4$ (326.4)	- 123° (pyridine)	From methanol when diluted with acetone, fine needles, m.p. 220°
Glycine diethylamide	$C_{22}H_{30}O_2N_4$ (382.5)	- 110° (pyridine)	From acetone, clear pentagonal plates, m.p. 187-188°
Glycine ethyl ester	$C_{20}H_{25}O_3N_3$ (355.4)	- 114° (pyridine)	From methylene chloride when diluted with ether, crystals, m.p. 200°
L-alanine methyl ester	$C_{20}H_{25}O_3N_3$ (355.4)	- 150° (pyridine)	From methylene chloride when diluted with ether, crystals, m.p. 220°
ß-Alanine methyl ester	$C_{20}H_{25}O_3N_3$ (355.4)	- 117° (pyridine)	From methylene chloride when diluted with ether, crystals, m.p. 173-175°
L-Leucine methyl ester	$C_{23}H_{31}O_3N_3$ (397.5)	- 139° (pyridine)	From ethyl acetate, crystals, m.p. 206-207°
L-Serine methyl ester	$C_{20}H_{25}O_4N_3$ (371.4)	- 110° (pyridine)	From ethyl acetate, long prisms, m.p. 197-198°
L-Proline ethyl ester	$C_{23}H_{29}O_3N_3$ (395.5)	- 104° (pyridine)	From benzene, crystals, m.p. 183-185°
L-Phenylalanine methyl ester	$C_{26}H_{29}O_3N_3$ (431.5)	- 104° (pyridine)	From acetone/benzene, crystals, m.p. 192-194°
L-Tryptophane ethyl ester	$C_{29}H_{32}O_3N_4$ (484.6)	- 72° (pyridine)	From ethyl acetate, crystals, m.p. 204-206°

Table 25 (Continuation)

Amino acid or peptide component	Empirical formula (Mol. wt.)	$[\alpha]_D^{20}$	Typical crystallization and m.p. (dec.)
Glycyl-L-leucine methyl ester	$C_{25}H_{34}O_4N_4$ (454.6)	- 97° (pyridine)	From acetone/ether, crystals, m.p. 120°
Diglycylglycine methyl ester	$C_{23}H_{29}O_5N_5$ (455.5)	- 85° (pyridine)	From methanol/acetone, crystals, m.p. 208°
D-Serine methyl ester	$C_{20}H_{25}O_4N_3$ (371.4)	- 108° (pyridine)	From ethyl acetate, needles, m.p. 192°
L-O-Acetyl-serine methyl ester	$C_{22}H_{27}O_5N_3$ (413.4)	- 114° (pyridine)	From CH$_2$Cl$_2$/ether, plates, mp. 160°
L-Serine amide	$C_{19}H_{24}O_3N_4$ (356.4)	- 102° (pyridine)	From diluted alcohol, needles, m.p. 225°
L-O-Acetyl-serine amide	$C_{21}H_{26}O_4N_4$ (398.4)	- 101° (pyridine)	From diluted alcohol, prisms, m.p. 194°
L-Seryl-L-leucine methyl ester	$C_{26}H_{36}O_5N_4$ (484.5)	- 99° (pyridine)	From ethyl acetate, prisms, m.p. 185°
D-Seryl-L-leucine methyl ester	$C_{26}H_{36}O_5N_4$ (484.5)	- 85° (pyridine)	From acetone, needles, m.p. 250°
L-Seryl-L-leucyl-L-proline methyl ester	$C_{31}H_{43}O_6N_5$ (581.7)	- 125° (pyridine)	From diluted alcohol, prisms, m.p. 131°
D-Seryl-L-leucyl-L-proline methyl ester	$C_{31}H_{43}O_6N_5$ (581.7)	- 104° (pyridine)	amorphous
L-Seryl-L-leucyl-D-proline	$C_{30}H_{41}O_6N_5$ (567.6)	- 71° (pyridine)	From alcohol, needles, m.p. 230°
Amino acid or peptide component	Empirical formula (Mol. wt.)	$[\alpha]_D^{20}$	Typical crystallization and m.p. (dec.)
D-Seryl-L-leucyl-D-proline methyl ester	$C_{31}H_{43}O_6N_5$ (581.7)	- 74° (pyridine)	From CH$_2$Cl$_2$/ether, needles, mp. 146°
L-Dimethylserine methyl ester	$C_{22}H_{29}O_4N_3$ (399.4)	- 114° (pyridine)	From benzene, needles, mp. 182°
D-Dimethylserine methyl ester	$C_{22}H_{29}O_4N_3$ (399.4)	- 105° (pyridine)	From acetone/benzene, prisms, mp. 216°
L-Dimethylserine amide	$C_{21}H_{28}O_3N_4$ (384.4)	- 113° (pyridine)	From dil. methanol, needles, mp. 154°
D-Dimethylserine amide	$C_{21}H_{28}O_3N_4$ (384.4)	- 92° (pyridine)	From methanol, prisms, mp. 233°
L-Dimethylseryl-L-leucine methyl ester	$C_{28}H_{40}O_5N_4$ (512.6)	- 102° (pyridine)	From dil. methanol, needles, mp. 195°
D-Dimethylseryl-L-leucine methyl ester	$C_{28}H_{40}O_5N_4$ (512.6)	- 79° (pyridine)	From acetone, needles, mp. 204°
D-Dimethylseryl-L-leucyl-D-proline methyl ester	$C_{28}H_{40}O_5N_4$ (609.7)	- 61° (pyridine)	amorphous
Dehydrovaline methyl ester	$C_{22}H_{27}O_3N_3$ (381.4)	+ 130° (pyridine)	From methanol, needles, m.p. 246°
Dehydrovaline amide	$C_{21}H_{26}O_2N_4$ (366.4)	- 126° (pyridine)	From methanol, prisms, m.p. 241°
Dehydrovaline methyl amide	$C_{22}H_{28}O_2N_4$ (380.5)	- 125° (pyridine)	From methanol, prisms, m.p. 274°

Table 25 (Continuation)

Dehydrovaline dimethyl amide	$C_{25}H_{34}O_2N_4$ (422.5)	-109° (pyridine)	From alcohol/ethyl acetate, leaflets, m.p. 234°
Dehydrovalyl-L-phenylalanine methyl ester	$C_{31}H_{36}O_4N_4$ (528.6)	-52° (pyridine)	From ethyl acetate, prisms, m.p. 239°
Dehydrovalyl-L-phenylalanyl-D-proline methyl ester	$C_{36}H_{43}O_5N_5$ (625.7)	-32° (pyridine)	From diluted acetone, prisms, m.p. 143°

Table 26

Peptides associated with dihydroisolysergic acid-(I) and with dihydroisolysergic acid-(II)
$C_{15}H_{17}N_2 \cdot COR$

Dihydroisolysergic acid-(I) Amino acid component	Empirical formula (Mol. wt.)	$[\alpha]_D^{20}$	Typical crystallization and m.p. (dec.)
Glycine amide	$C_{18}H_{22}O_2N_4$ (326.4)	+16° (pyridine)	From acetone, fine needles, m.p. 225°
Glycine ethyl ester	$C_{20}H_{25}O_3N_3$ (355.4)	+23° (pyridine)	From ether, prisms, m.p. 95°
Dihydroisolysergic acid-(II) Amino acid component			
Glycine amide	$C_{18}H_{22}O_2N_4$ (326.4)	+41° (pyridine)	Slightly soluble in acetone, sparingly soluble in benzene. Did not crystallize
Glycine ethyl ester	$C_{20}H_{25}O_3N_3$ (355.4)	+23° (pyridine)	From alcohol or acetone, leaflets, m.p. 175°

According to the method of *W. L. Garbrecht*[242], the lithium salt of lysergic acid in dimethylformamide solution is reacted with SO_3 to give the mixed lysergic acid-sulfuric anhydride, which reacts with primary or secondary amines in good yield to give the corresponding lysergic acid amides.

The method of *R. P. Pioch*[243] also leads via a mixed anhydride, namely the mixed lysergic acid-trifluoroacetic acid anhydride. However, the yields of lysergic acid amide here are usually not as good as in the SO_3 procedure.

Another technically feasible route for the preparation of lysergic acid amides, which has been followed in the Sandoz laboratories, uses lysergic acid chloride hydrochloride as activated form[100, 170]. This reacts at low temperature with an excess of primary or secondary amine to give the corresponding lysergic acid amide and amine hydrochloride.

Furthermore, the method developed for peptide syntheses using *N,N'*-carbonyldiimidazole as a condensing agent[244] can also be used to prepare acid amides of lysergic acid and dihydrolysergic acid[245].

A further process for the preparation of amide-like derivatives of lysergic acid and dihydrolysergic acid consists in the use of dialkyl pyrazoles, which are formed from the lysergic acid or dihydrolysergic acid hydrazides on reaction with diketones of the formula R'-COCH$_2$COR", as activated intermediates[385].

2. Amino and carbamic acid derivatives of 6-methylergolenes and 6-methylergolines

Starting from d-lysergic acid azide or d-isolysergic acid azide[157] and from the isomeric d-dihydrolysergic acid azides[121], the isomeric 6-methyl-8-amino-ergolenes (Ia) and isomeric 6-methyl-8-amino-ergolines (IIa) have been prepared by a modified Curtius degradation[131] (see Formula Scheme 25). The process consists of briefly boiling the azide hydrochlorides with dilute aqueous hydrochloric acid, whereby under CO$_2$ evolution the corresponding amine hydrochlorides are directly formed. Table 27 shows the data for the isomeric 6-methyl-8-amino- ergolene and -ergoline.

Formula scheme 25

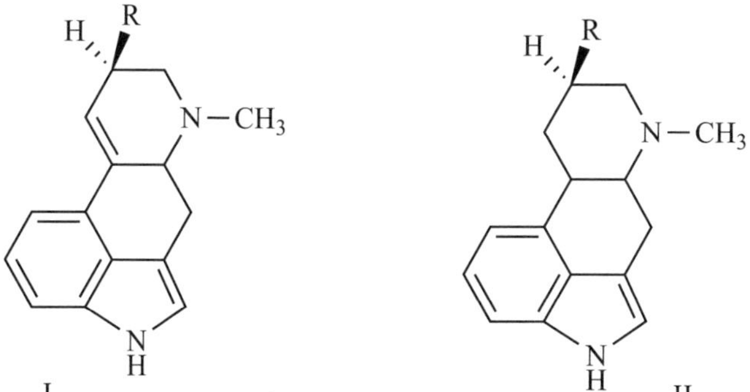

a) : R = NH$_2$

b) : R = NHCOOR'

c) : R = NHCON$\diagup\!\!\diagdown$ $\overset{\text{R'}}{\underset{\text{R''}}{}}$

On reaction of the isocyanates, formed on heating of the lysergic or isolysergic azide in benzene, with alcohols R'OH the corresponding carbamic acid esters Ib and IIb are obtained (Formula Scheme 25)[132]. The properties of a number of homologous urethanes are shown in Table 28.

163

Table 27

Properties of the isomeric 6-methyl-8-amino-ergolenes, $C_{15}H_{17}N_3$ (239.3)
and 6-methyl-8-amino-ergolines, $C_{15}H_{19}N_3$ (241.3)

	m.p. (with dec.)	$[\alpha]_D^{20}$ (pyridine)	Typical crystallization
6-Methyl-8-amino-ergolene	253°	+96°	From chloroform, in elongated, angular plates
6-Methyl-8-amino-iso-ergolene	198°	+249°	From alcohol, in solid prisms and polyhedra
6-Methyl-8-amino-ergoline	243°	−117°	From ethyl acetate, in solid, obliquely cut prisms
6-Methyl-8-amino-iso-ergoline-(I)	275-280° (Block)	−66°	From methanol, in elongated, hexagonal sheets
6-Methyl-8-amino-iso-ergoline-(II)	203°	+29°	From ethyl acetate, in solid straight cut prisms

Table 28

Homologous esters of (6-methyl-ergolen-8-yl)-
and (6-methyl-isoergolen-8-yl)-carbamic acid

	m.p. (with dec.)	$[\alpha]_D^{20}$ (pyridine)	Typical crystallization
(6-Methyl-ergolen-8-yl)-carbamic acid methyl ester $C_{17}H_{19}O_2N_3$ (297.3)	236-237°	+50°	From methanol, prisms sharpened on both sides, from chloroform pentagonal leaflets
(6-Methyl-ergolen-8-yl)-carbamic acid ethyl ester $C_{18}H_{23}O_2N_3$ (311.4)	237-238°	+48°	From benzene, 6- to 8-conrered leaflets
(6-Methyl-ergolen-8-yl)-carbamic acid propyl ester $C_{19}H_{23}O_2N_3$ (325.4)	200-202°	+47°	From benzene, hexagonal plates, from alcohol sheets
(6-Methyl-ergolen-8-yl)-carbamic acid butyl ester $C_{20}H_{25}O_2N_3$ (399.4)	207-208°	+42°	From benzene, fine needles, from alcohol large 5- to 8-cornered plates
(6-Methyl-isoergolen-8-yl)-carbamic acid methyl ester $C_{17}H_{19}O_2N_3$ (297.3)	180°	+346°	From alcohol, polyhedra
(6-Methyl-isoergolen-8-yl)-carbamic acid ethyl ester $C_{18}H_{23}O_2N_3$ (311.4)	177°	+326°	From benzene and alcohol, long, straight cut prisms
(6-Methyl-isoergolen-8-yl)-carbamic acid propyl ester $C_{19}H_{23}O_2N_3$ (325.4)	—	+294°	amoprhous
(6-Methyl-isoergolen-8-yl)-carbamic acid butyl ester $C_{20}H_{25}O_2N_3$ (399.4)	—	+270°	amorphous

Table 29

N-(6-methyl-ergolin-8-yl)-N'-cycloalkyl ureas

Substance	Empirical formula (Mol. wt.)	m.p. (dec.)	$[\alpha]_D^{20}$ (pyridine)
N-(6-Methyl-ergolin-8-yl)-N'-			
(cyclopropyl)urea	$C_{19}H_{24}ON_4$ (324.4)	>310°	−72°
(cyclobutyl)urea	$C_{20}H_{26}ON_4$ (338.4)	>310°	−75°
(cyclopentyl)urea	$C_{21}H_{28}ON_4$ (352.4)	>310°	−70°
(cyclohexyl)urea	$C_{22}H_{30}ON4$ (366.5)	>310°	−71°
(cycloheptyl)urea	$C_{23}H_{32}ON_4$ (380.5)	>310°	−67°
(cyclopentylpropyl)urea	$C_{24}H_{34}ON_4$ (394.5)	>310°	−63°

Table 30

Urea with the 6-methyl-ergolinyl, 6-methyl-ergolenyl and 6-methyl-isoergolenyl scaffold

Substance	Empirical formula (Mol. wt.)	m.p. (dec.)	$[\alpha]_D^{20}$ (pyridine)
6-Methyl-ergolin-8-yl-NHCOR			
R 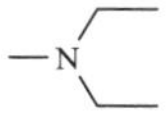 = —N< (azetidine)	$C_{20}H_{28}ON_4$ (340.5)	260°	−55°
R 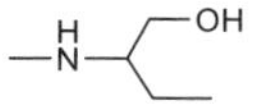 = —NH—CH(CH₃)—OH	$C_{20}H_{28}O_2N_4$ (356.6)	260°	−95°
R 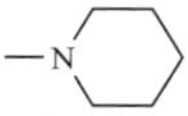 = —N< (piperidine)	$C_{21}H_{28}ON_4$ (352.5)	260°	−55°
6-Methyl-ergolin-8-yl-NHCOR			
R 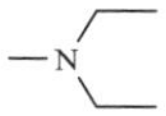 = —N< (azetidine)	$C_{20}H_{26}ON_4$ (338.4)	242°	+112°
R 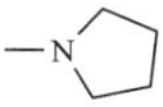 = —N< (pyrrolidine)	$C_{21}H_{26}ON_4$ (350.5)	247°	+63°
R 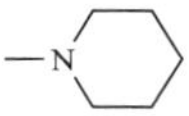 = —N< (piperidine)	$C_{21}H_{26}ON_4$ (350.5)	228°	+104.5°
6-Methyl-isoergolen-8-yl-NHCOR			
R 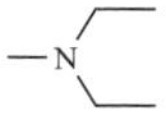 = —N< (azetidine)	$C_{20}H_{26}ON_4$ (338.4)	186°	+313°
R 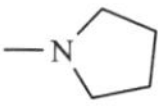 = —N< (pyrrolidine)	$C_{21}H_{26}ON_4$ (350.5)	152°	+317°
R 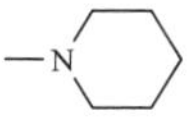 = —N< (piperidine)	$C_{21}H_{26}ON_4$ (350.5)	115°	+297°

Ergolene and ergoline derivatives substituted on the indole nitrogen with a urethane grouping are also described in the patent literature[246].

Reaction of isocyanates with amines $HN\langle^{R'}_{R''}$ leads to urea derivatives with of the structure Ic and IIc (Formula Scheme 25)[247, 248]. A number of such urea derivatives with the ring skeleton of the ergot alkaloids are compiled in Tables 29 and 30.

Further ergolene and ergoline derivatives substituted with the urea moiety are described in the patent literature[249, 250].

In pharmacological testing of these urea derivatives, N-(6-methyl-isoergolen-8-yl)-N',N'-(diethyl)urea (lysenyl) proved to be a highly potent serotonin antagonist[251, 252]. It is noteworthy that in this case, as an exception, the isoergolenyl derivative is more active than the ergolenyl compound.

3. Substitutions on the ring system of lysergic acid

On the ring system of lysergic acid, positions 1 and 2 are particularly reactive, at which various substituents can be introduced.

a) Substitutions on indole nitrogen

In derivatives of lysergic acid and dihydrolysergic acid in which the carboxyl group is substituted in the manner of an ester or acid amide, the indole nitrogen can be acetylated with ketene (see Formula Scheme 26a and b), substituted with formaldehyde by the hydroxymethyl residue c), or by means of a Mannich reaction by the dialkylaminomethyl residue e)[253].

Formula Scheme 26

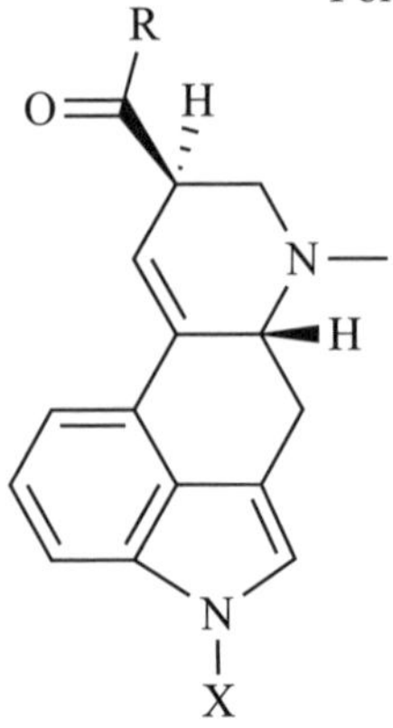

a) X = COCH$_3$
b) X = COCH$_2$COCH$_3$
c) X = CH$_2$OH
d) X = CH$_2$OCOCH$_3$
e) X = CH$_2$N(alkyl)$_2$
f) X = Alkyl or benzyl

The hydrogen on the indole nitrogen can also be alkylated (f)[254, 255]. The acetylation reaction can be carried out in benzene solution with the addition of a catalytic amount of triethylamine. If acetone is used as a solvent, the addition of a basic catalyst is unnecessary. In this medium, however, a larger amount of the acetoacetyl derivative b) is formed in addition to the monoacetyl compound a). In the case of the peptide alkaloids, additional transformations occur with the ketene, so that no uniform reaction products could be obtained here. The N-acetyl bond is stable in moderately acidic and in bicarbonate-alkaline environments. With sodium carbonate it is hydrolytically cleaved on heating, with alkali already at room temperature. Table 31 shows the data for some 1-acetyl and 1-acetoacetyl derivatives of lysergic and dihydrolysergic acid compounds.

Table 31
1-Acyl derivatives of lysergic and dihydrolysergic acid compounds

1-Acetyl derivative of	Empirical formula (Mol. wt.)	$[\alpha]_D^{20}$ Chl = in CHCl$_3$ Py = in pyridine	m.p. (dec.)
d-Lysergic acid methyl ester	C$_{19}$H$_{20}$O$_3$N$_2$ (324.4)	+ 15° (Chl)	179-181°
d-Lysergic acid amide	C$_{18}$H$_{19}$O$_2$N$_3$ (309.4)	+ 46° (Py)	226°
d-Isolysergic acid amide	C$_{18}$H$_{19}$O$_2$N$_3$ (309.4)	+ 370° (Py)	150-154°
d-Lysergic acid ethylamide	C$_{20}$H$_{23}$O$_2$N$_3$ (337.4)	+ 40° (Py)	218-220°
d-Lysergic acid diethylamide	C$_{22}$H$_{27}$O$_2$N$_3$ (365.5)	+ 14° (Py)	amorphous
d-O-acetylergometrine	C$_{23}$H$_{27}$O$_4$N$_3$ (409.5)	+ 59° (Py)	185-192°
d-Dihydrolysergic acid-(I) methyl ester	C$_{19}$H$_{22}$O$_3$N$_2$ (326.4)	+ 98° (Py)	182-183°
d-Dihydrolysergic acid-(I) diethyl amide	C$_{23}$H$_{29}$O$_2$N$_3$ (367.5)	+ 104° (Py)	174-175°
d-(6-Methyl-isoergolenyl-8)-carbamic acid ethyl ester	C$_{20}$H$_{23}$O$_3$N$_3$ (353.4)	+ 287° (Chl)	156-159°
1-Acetoacetyl derivative of			
d-Lysergic acid methyl ester	C$_{21}$H$_{22}$O$_4$N$_2$ (366.4)	− 8° (Chl)	168-169°
d-Dihydrolysergic acid-(I) methyl ester	C$_{21}$H$_{24}$O$_4$N$_2$ (368.4)	− 114° (Py)	190-191°

For hydroxymethylation, an acetic acid solution of the lysergic acid derivative is heated with excess formaldehyde solution to 60–80°. The reaction occurs only in acidic solution. On the other hand, the acetic acid concentration must not exceed 25%, otherwise, besides the hydroxymethyl compound c) also the corresponding acetoxymethyl derivative d) is formed.

Table 32

1-Hydroxymethyl and 1-acetoxymethyl derivatives of lysergic acid dihydrolysergic acid compounds

1-Hydroxymethyl derivative of	Empirical formula (Mol. wt.)	$[\alpha]_D^{20}$ Chl = in $CHCl_3$ Py = in pyridine	m.p. (dec.)
d-Lysergic acid diethylamide	$C_{21}H_{27}O_2N_2$ (353.5)	+23° (Py)	164-166°
Ergobasine	$C_{20}H_{25}O_3N_3$ (355.4)	+11° (Py)	125-126°
Ergotamine	$C_{34}H_{37}O_6N_3$ (611.7)	+131° (Cl)	amorphous
Ergotaminine	$C_{34}H_{37}O_6N_3$ (611.7)	+335° (Cl)	201-204°
d-Dihydrolysergic acid-(I) methyl ester	$C_{18}H_{22}O_3N_2$ (314.4)	+100° (Py)	191-192°
Dihydroergotamine	$C_{34}H_{39}O_6N_5$ (613.7)	+52° (Py)	amorphous
Dihydroergocristine	$C_{36}H_{43}O_6N_5$ (641.7)	+55° (Py)	amorphous
Dihydroergocornine	$C_{32}H_{43}O_6N_5$ (593.7)	+43° (Py)	179-180°
Dihydroergocryptine	$C_{33}H_{45}O_6N_5$ (607.7)	+41° (Py)	amorphous
1-Acetoxymethyl derivative of			
d-Dihydrolysergic acid-(I) methyl ester	$C_{20}H_{24}O_4N_2$ (356.4)	−91° (Py)	113-114°
Dihydroergocristine	$C_{38}H_{45}O_7N_5$ (683.8)	+51° (Py)	199-200°
Dihydroergocornine	$C_{34}H_{45}O_7N_5$ (635.7)	+35° (Py)	168-170°
Dihydroergocryptine	$C_{35}H_{47}O_7N_5$ (649.8)	+30° (Py)	223-224°

Table 33
Mannich bases of lysergic acid and dihydrolysergic acid compounds

1-Dimethylaminomethyl derivative of	Empirical formula (Mol. wt.)	$[\alpha]_D^{20}$ in pyridine	m.p. (dec.)
d-Lysergic acid diethylamide	$C_{23}H_{32}ON_4$ (365.5)	+15°	amoprhous
d-Dihydrolysergic acid-(I) methyl ester	$C_{20}H_{27}O_2N_3$ (341.4)	+93°	93-94°
Dihydroergotamine	$C_{38}H_{44}O_5N_6$ (640.8)	+62°	165-172°
Dihydroergocristine	$C_{38}H_{48}O_5N_6$ (668.8)	+29°	189-190°
Dihydroergocornine	$C_{34}H_{45}O_7N_5$ (620.8)	+34°	207°
Dihydroergocryptine	$C_{35}H_{50}O_5N_6$ (634.8)	+31°	183-187°
1-Acetoacetyl derivative of			
d-Dihydrolysergic acid-(I) methyl ester	$C_{23}H_{31}O_2N_3$ (381.5)	+85°	89-91°

Like the acetyl group, the hydroxymethyl group is stable to hydrogen carbonate, but is also hydrolytically cleaved by soda when heated and by dilute alkali when cold. Table 32 contains the data for a number of hydroxymethyl and acetoxymethyl derivatives.

The 1-dialkylaminomethyl derivatives of the lysergic acid and dihydrolysergic acid series prepared by Mannich reaction are summarized in Table 33. They are stable at room temperature both against moderately strong alkali and against dilute acids, but split off the dialkylaminomethyl group again on heating.

For the 1-alkylation of lysergic acid and dihydrolysergic acid derivatives, these are treated in liquid ammonia with the addition of 1 equivalent of potassium amide with the alkyl iodide or bromide in question. The alkylation proceeds particularly smoothly on the free lysergic acid[245, 255].

In the case of lysergic acid compounds, alkylation also occurs at C8 with an excess of potassium amide and methyl iodide[254]. Table 34 contains the most important data for a number of 1-alkyl derivatives of lysergic acid and dihydrolysergic acid compounds.

While the introduction of the easily cleavable acyl or hydroxymethyl residues does not significantly alter the pharmacological properties of the starting materials, and the corresponding Mannich bases also do not represent an interesting modification, the substitution of the indole nitrogen with alkyl changes the pharmacodynamic character of the lysergic acid and dihydrolysergic acid derivatives in a specific way. In particular, the serotonin antagonistic effect of the parent compounds is substantially enhanced. Of the 1-alkyl derivatives listed in Table 34, 1-methyl-*d*-lysergic acid (+)-butanolamide-(2) has achieved medical importance. The compound is a highly active serotonin antagonist and is used under the brand name "Deseril"® as a drug for the treatment of migraine (see Section D).

Table 34
1-Alkyl derivatives of lysergic acid and dihydrolysergic acid compounds

Substance	Empirical formula (Mol. wt.)	$[\alpha]_D^{20}$ (pyridine)	m.p. (dec.)
1-Methyl derivative of			
d-Lysergic acid	$C_{17}H_{18}O_2N_2$ (282.4)	+120° (0.1 N CH_3SO_3H)	237-239°
d-Isolysergic acid	$C_{17}H_{18}O_2N_2$ (282.4)	+330°	215°
d-Lysergic acid amide	$C_{17}H_{19}ON_3$ (281.4)	−5°	190-192°
d-Isolysergic acid amide	$C_{17}H_{19}ON_3$ (281.4)	+449°	197-198°
d-Lysergic acid ethylamide	$C_{19}H_{23}ON_3$ (309.4)	−4°	185-186°
d-Lysergic acid diethylamide	$C_{21}H_{27}ON_3$ (337.5)	+20°	amorphous
d-Lysergic acid pyrrolidide	$C_{21}H_{25}ON_3$ (337.5)	+27°	amorphous
d-Lysergic acid L-propanolamide-(2)	$C_{20}H_{25}O_2N_3$ (339.4)	−22°	178-179°
d-Lysergic acid (+)-butanolamide-(2)	$C_{21}H_{27}O_2N_3$ (353.4)	−66°	181-186°
Ergotamine	$C_{34}H_{37}O_5N_5$ (595.7)	−170° (in $CHCl_3$)	185°
Ergotaminine	$C_{34}H_{37}O_5N_5$ (595.7)	+403	224°

Dihydro-*d*-lysergic acid	$C_{17}H_{20}O_2N_2$ (284.4)	$-111°$	235°
Dihydro-*d*-lysergic acid methyl ester	$C_{18}H_{22}O_2N_2$ (298.4)	$-99°$	117-118°
Dihydro-*d*-lysergic acid amide	$C_{17}H_{21}ON_3$ (283.4)	$-135°$	256°
Dihydro-*d*-lysergic acid diethylamide	$C_{21}H_{29}ON_3$ (339.5)	$-111°$	131-132°
Dihydroergotamine	$C_{34}H_{39}O_5N_5$ (597.7)	$-67°$	224°
Dihydroergocristine	$C_{36}H_{43}O_5N_5$ (625.7)	$-57°$	225°
Dihydroergocryptine	$C_{33}H_{45}O_5N_5$ (591.7)	$-40°$	244-245°
Dihydroergocornine	$C_{32}H_{43}O_5N_5$ (577.7)	$-44°$	215-218°
1-Ethyl derivative of			
d-Lysergic acid	$C_{18}H_{22}O_2N_2$ (296.4)	$+113°$ (0.1 N CH$_3$SO$_3$H)	219-220°
Dihydroergocornine	$C_{33}H_{45}O_5N_5$ (591.7)	$-44°$	176-177°
1-Propyl deriviative of			
d-Lysergic acid	$C_{18}H_{22}O_2N_2$ (296.4)	$+102°$ (0.1 N CH$_3$SO$_3$H)	206-208°
Dihydroergocornine	$C_{34}H_{47}O_5N_5$ (605.8)	$-47°$	184-185°
1-Allyl derivative of			
d-Lysergic acid	$C_{19}H_{20}O_2N_2$ (308.4)	$+99°$ (0.1 N CH$_3$SO$_3$H)	209-211°
Dihydroergocristine	$C_{38}H_{45}O_5N_5$ (651.8)	$-51°$	166-168°
Dihydroergocryptine	$C_{38}H_{47}O_5N_5$ (617.8)	$-40°$	224-225°
Dihydroergocornine	$C_{31}H_{45}O_5N_5$ (603.7)	$-50°$	180-181°
1-Benzyl derivative of			
Dihydro-*d*-lysergic acid	$C_{23}H_{24}O_2N_2$ (360.4)	$-106°$	217-222°
Dihydroergocristine	$C_{42}H_{47}O_5N_5$ (701.8)	$-58°$	155-160°
Dihydroergocryptine	$C_{39}H_{49}O_5N_5$ (667.8)	$-51°$	154-160°
Dihydroergocornine	$C_{38}H_{47}O_5N_5$ (653.8)	$-51°$	170-174°

b) Halogenation in the 2-position

Derivatives of lysergic acid with the 1-position free or occupied by an alkyl residue can be halogenated in the 2-position with bromo- or iodosuccinimide, N-chloro-N-(2,6-dichloro-4-nitrophenyl)acetamide, and similar mild-acting reagents (Formula Scheme 27)[256].

Formula Scheme 27

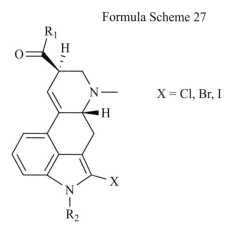

X = Cl, Br, I

Table 35
2-Chloro-lysergic acid derivatives

Substance	Empirical formula (Mol. wt.)	$[\alpha]_D^{20}$ Chl = in CHCl₃ Py = in pyridine	Typical crystallization and m.p. (dec.)
2-Chloro-ergotamine	$C_{33}H_{34}O_5N_5Cl$ (616.1)	- 157° (Chl)	From 90% acetone rectangular plates, m.p. 170-172°
2-Chloro-dihydro-d-lysergic acid-(I) methyl ester	$C_{17}H_{19}O_2N_2Cl$ (318.8)	- 119° (Py)	From benzene needles, m.p. 209-211°
2-Chloro-dihydroergotamine	$C_{33}H_{36}O_5N_5Cl$ (618.1)	- 80° (Py)	From chloroform prisms, m.p. 176⁄178°
2-Chloro-dihydroergocristine	$C_{35}H_{40}O_5N_5Cl$ (646.2)	- 70° (Py)	From from benzene oblique prisms, m.p. 172-175°

The halogen atom in the 2-position of the lysergic acid derivatives is extremely inert and could not be exchanged for other groups, e.g., by the action of alkali, alcoholates, or organic bases, even under vigorous conditions[272]. On the other hand, the halogen can be removed reductively in an alkaline environment. The van Urk color reaction, which requires a free 2-position, is negative for the halogen derivatives of lysergic acid and dihydrolysergic acid. Tables 35–38 list the most important data for chloro-, bromo-, and iodo-lysergic acid derivatives.

Table 36
2-Bromo-lysergic acid derivatives

Substance	Empirical formula (Mol. wt.)	$[\alpha]_D^{20}$ Chl = in CHCl$_3$ Py = in pyridine	Typical crystallization and m.p. (dec.)
2-Bromo-*d*-lysergic acid methyl ester	C$_{17}$H$_{17}$O$_2$N$_2$Br (361.3)	+41° (Chl)	From beznene, short prisms, m.p. 177-178°
2-Bromo-*d*-lysergic acid amide	C$_{16}$H$_{16}$ON$_3$Br (346.2)	+15° (Py)	From ethyl acetate, prisms, m.p. 200-202°
2-Bromo-*d*-isolysergic acid amide	C$_{16}$H$_{16}$ON$_3$Br (346.2)	+454° (Py)	From chloroform, prisms, m.p. 215-217°
2-Bromo-*d*-lysergic acid ethylamide	C$_{18}$H$_{20}$ON$_3$Br (374.3)	−2° (Py)	From from benzene, double pyramids, m.p. 128-129°
2-Bromo-*d*-lysergic acid pyrrolidide bitartrate	C$_{20}$H$_{22}$ON$_3$Br, C$_4$H$_{10}$O$_6$ (550.4)	+37° (Chl)	Base amorphous. Acid tartrate from acetone, needles, m.p. 165-170°
2-Bromo-*d*-lysergic acid diethylamide	C$_{20}$H$_{24}$ON$_3$Br (402.3)	+15° (Py)	From ether, needles, m.p. 120-127°
2-Bromo-ergotamine	C$_{33}$H$_{34}$O$_5$N$_5$Br (660.6)	−163° (Chl)	From 90%. acetone, rectangular plates, m.p. 195-197°
2-Bromo-ergocristine	C$_{35}$H$_{38}$O$_5$N$_5$Br (688.6)	−189° (Chl)	From benzene, spherical aggregates, m.p. 183-187°
2-Bromo-ergocornine	C$_{31}$H$_{38}$O$_5$N$_5$Br (640.6)	−215° (Chl)	From chloroform, prisms sharpened on both sides, m.p. 187-193°
2-Bromo-ergocorninine	C$_{31}$H$_{38}$O$_5$N$_5$Br (640.6)	+425° (Chl)	From methanol, needles, m.p. 190-198°
2-Bromo-ergobasine	C$_{19}$H$_{22}$O$_2$N$_3$Br (404.3)	−15° (Py)	From chloroform, rods, m.p. 110-128°
2-Bromo-dihydro-*d*-lysergic acid-(I)	C$_{16}$H$_{17}$O$_2$N$_2$Br, H$_2$O (367.3)	−156° (Py)	From water, spikes, m.p. 295-297°
2-Bromo-dihydro-*d*-lysergic acid-(I) methyl ester	C$_{17}$H$_{19}$O$_2$N$_2$Br (363.3)	−135° (Py)	From methanol, prisms, m.p. 215°
2-Bromo-dihydro-*d*-lysergic acid-(I) hydrazide	C$_{16}$H$_{19}$ON$_4$Br (363.3)	−134° (Py)	From chloroform, spikes, m.p. 257-259°
2-Bromo-dihydro-*d*-lysergic acid-(I) amide	C$_{16}$H$_{18}$ON$_3$Br (348.2)	−148° (Py)	Sublimed in high vacuum at 220-230°, m.p. 256°
2-Bromo-dihydro-*d*-lysergyl glycinamide	C$_{18}$H$_{21}$O$_2$N$_4$Br (405.3)	−125° (Py)	From alcohol, needles, m.p. 242-246°
2-Bromo-dihydroergotamine	C$_{33}$H$_{36}$O$_5$N$_5$Br (662.6)	−87° (Py)	From 90% acetone, rectangular plates, m.p. 198-199°
2-Bromo-dihydroergocristine	C$_{35}$H$_{40}$O$_5$N$_5$Br (690.6)	−63° (Py)	From benzene, prisms and plates, m.p. 184-185°
2-Bromo-dihydroergocryptine	C$_{32}$H$_{42}$O$_5$N$_5$Br (656.7)	−57° (Py)	From benzene, rods, m.p. 188-190°
2-Bromo-dihydroergocornine	C$_{31}$H$_{40}$O$_5$N$_5$Br (642.6)	−59° (Py)	From chloroform, needles, m.p. 186-189°

Substitution of the 2-position with halogen largely alters the pharmacological properties of the parent substances. Thus, 2-bromo-lysergic acid diethylamide lacks the hallucinogenic effect that characterizes lysergic acid diethylamide (LSD-25), while the serotonin-blocking effect of this compound is enhanced by bromination[257-260] (cf. also Section D).

Table 37
2-Iodo-lysergic acid derivatives

Substance	Empirical formula (Mol. wt.)	$[\alpha]_D^{20}$ Chl = CHCl$_3$ Py = pyridine	Typical crystallization and m.p. (dec.)
2-Iodo-d-lysergic acid diethylamide	C$_{20}$H$_{24}$ON$_3$I (449.3)	+ 22° (Py)	From beznene, needles, m.p. 133-138° or polyhedra, m.p. 206-208°
2-Iodo-ergotamine	C$_{33}$H$_{34}$O$_5$N$_5$I (707.6)	- 156° (Chl)	From 90% acetone, rectangular plates, m.p. 174°
2-Iodo-dihydro-d-lysergic acid-(I) methyl ester	C$_{17}$H$_{19}$O$_2$N$_2$I (410.3)	- 130° (Py)	From methanol or chloroform, needles, m.p. 247-249°
2-Iodo-dihydroergotamine	C$_{33}$H$_{34}$O$_5$N$_5$I (709.6)	- 90° (Py)	From 90% acetone, rectangular plates, m.p. 179-181°
2-Iodo-dihydroergocristine	C$_{33}$H$_{40}$O$_5$N$_5$I (737.6)	- 82° (Py)	From benzene, hexagonal sheets or irregular prisms, m.p. 187-192°

Table 38
1-Alkyl-2-halogen-lysergic acid derivatives

Substance	Empirical formula (Mol. wt.)	$[\alpha]_D^{20}$ Py = pyridine	Typical crystallization and m.p. (dec.)
1-Methyl-2-bromo-d-lysergic acid diethylamide	C$_{21}$H$_{26}$ON$_3$Br (416.4)	+ 31° (Py)	Base amorphous. Neutral tartrate form acetone, needles, m.p. 189-190°
1-Methyl-2-bromo-dihydro-d-lysergic acid-(I) diethylamide	C$_{21}$H$_{28}$ON$_3$Br (418.4)	- 126° (Py)	From ethyl acetate, hexagonal plates, m.p. 138-139°
1-Methyl-2-bromo-dihydro-ergotamine	C$_{34}$H$_{38}$O$_5$N$_5$Br (676.6)	- 79° (Py)	From ethyl acetate, prisms, m.p. 232°
1-Methyl-2-bromo-dihydro-ergocryptine	C$_{33}$H$_{44}$O$_5$N$_5$Br, C$_6$H$_6$ (748.7)	- 54° (Py)	From benzene, spherical crystal aggregates, m.p. 230-231°
1-Methyl-2-bromo-dihydro-ergocornine	C$_{32}$H$_{42}$O$_5$N$_5$Br (656.6)	- 55° (Py)	amorphous
1-Propyl-2-bromo-dihydro-ergocornine	C$_{34}$H$_{46}$O$_5$N$_5$Br (684.7)	- 61° (Py)	From alcohol, plates, m.p. 222-223°
1-Methyl-2-iodo-dihydro-ergotamine	C$_{34}$H$_{38}$O$_5$N$_5$I (723.6)	- 84° (Py)	From ethyl acetate, prisms, m.p. 222-223°

c) Saturation of the double bond in the 9,10-position

1) Dihydro derivatives

The saturation of the carbon double bond in the 9,10-position of the ergolene system by catalytic hydrogenation, which has already been discussed in Section C, III/2 d concerning the stereochemistry of the reaction products, led in the series of peptide alkaloids to medically valuable dihydro derivatives[120], which have found their way into therapy in the form of pharmaceutical preparations such as "Hydergine"® (mixture of equal parts of dihydroergocristine, dihydroergocryptine, and dihydroer-

gocornine in the form of their methane sulfonates) and "Dihydergot"®
(methane sulfonate of dihydroergotamine). Hydrogenation attenuates
uterotonic activity while greatly enhancing sympatholytic activity. Table
39 summarizes the data of the dihydro derivatives of the natural ergot
alkaloids[120] of the lysergic acid series and Table 40 of the isolysergic acid
series[121].

Table 39
The dihydro derivatives of the natural levorotatory ergot alkaloids

Substance	Gross formula (Mol. wt.)	$[\alpha]_D^{20}$ (pyridine)	Typical crystallization and m.p. (dec.)
Ergotamine group:			
Dihydroergotamine	$C_{33}H_{37}O_5N_5$ (583.7)	$-64°$	From 90% aqueous acetone, straight prisms with a large surface area, m.p. 239°
Dihydroergosine	$C_{30}H_{39}O_5N_5$ (549.7)	$-52°$	From ethyl acetate, sharpened prisms and polyhedra, m.p. 212°
Ergotoxine group:			
Dihydroergocristine	$C_{35}H_{41}O_5N_5$ (611.7)	$-56°$	From acetone, solid hexagonal plates, m.p. 180°
Dihydroergocryptine	$C_{32}H_{43}O_5N_5$ (577.7)	$-41°$	From ethyl alcohol, solid plates and polyhedra, m.p. 235°
Dihydroergocornine	$C_{31}H_{41}O_5N_5$ (563.7)	$-48°$	From ethyl alcohol, solid, mostly hexagonal plates, m.p. 230°
Ergobasine group:			
Dihydroergobasine	$C_{19}H_{25}O_2N_3$ (327.4)	$-145°$	From benzene, needles, m.p. 230°

Table 40
The dihydro derivatives of the dextrorotatory ergot alkaloids

Substance	Empirical formula (Mol. wt.)	m.p. (dec.)	$[\alpha]_D^{20}$ (pyridine)	Typical crystallization and m.p. (dec.)
Dihydroergotaminine-(I)	$C_{33}H_{37}O_5N_5$ (583.7)	236°	$+97°$	From alcohol, in hexagonal leaflets
Dihydroergotaminine-(II)		206°	$-7°$	From acetone, in rectangluar plates
Dihydroergosinine-(I)	$C_{30}H_{39}O_5N_5$ (549.7)	234°	$+108°$	From alcohol, in long needles
Dihydroergosinine-(II)		223°	$+3°$	From acetone, in fine needles
Dihydroergocristinine-(I)	$C_{35}H_{41}O_5N_5$ (611.7)	248°	$+109°$	From alcohol, in polyhedra
Dihydroergocristinine-(II)		175°	$+13°$	From acetone, in 6- or 8-angular plates
Dihydroergocryptinine-(I)	$C_{32}H_{43}O_5N_5$ (577.7)	268°	$+126°$	From methylene chloride-methanol, in long prisms
Dihydroergocryptinine-(II)		226°	$+26°$	From methylene chloride-petroleum ether, in spear-shaped needles
Dihydroergocorninine-(I)	$C_{31}H_{41}O_5N_5$ (563.7)	264°	$+147°$	From chloroform-alcohol, in long needles
Dihydroergocorninine-(II)		180°	$+32°$	From methylene chloride, in solid hexagonal plates
Dihydroergobasinine-(I)	$C_{19}H_{25}O_2N_3$ (327.4)	211°	$+8°$	From ethyl acetate-methanol, in leaflets
Dihydroergobasinine-(II)		212°	$+45°$	From acetone, in solid, obliquely cut prisms

2) Lumi derivatives

Water can also be added to the double bond in the 9,10-position. This reaction takes place in an acidic aqueous solution of the alkaloids when irradiated intensively with ultraviolet light. The water addition products formed under the influence of light became signified with the prefix "lumi"[204]. The newly entered hydroxyl group shows tertiary character, thus must be located at the C10. Based on the rotational values, the preferential formation tendency and the chromatographic behavior, the lumi isomers could be configurationally linked with the corresponding dihydrolysergic acid and dihydroisolysergic acid derivatives. The steric ratios with the absolute configurations are represented by the formulas in Scheme 28.

Compared with the spectrum of the corresponding lysergic acid derivatives, the UV spectrum of the lumi compounds shows the same shift of the absorption bands to the shorter-wavelength region as that of the dihydro compounds (cf. Figure 19).

Lumi compounds lack the characteristic fluorescent color of lysergic acid derivatives. All lumi derivatives dissolve in conc. sulfuric acid to form a deep violet-blue color. This color reaction allows them to be easily distinguished from the lysergic acid and dihydrolysergic acid compounds.

Formula Scheme 28

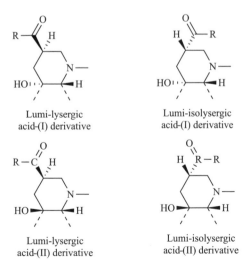

Lumi-lysergic
acid-(I) derivative

Lumi-isolysergic
acid-(I) derivative

Lumi-lysergic
acid-(II) derivative

Lumi-isolysergic
acid-(II) derivative

The lumi-alkaloids can be hydrolytically cleaved to the corresponding lumi-lysergic acids by boiling with dilute alkali solution.

Table 41 and 42 summarize the properties of the described lumi derivatives[204, 262–266]

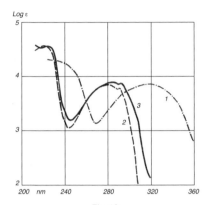

Fig. 19

Curve 1: — . — . — Ergotamine   Maximum at 318 mµ (3.86)
Curve 2: — — —   Dihydroergotamine   Maxima: 220 (4.54), 282 (3.85), 291 (3.78)
Curve 3: ————   Lumi-ergotamine   Maxima: 223 (4.54), 285 (3.88), 293 (3.87)

Table 41
Lumi derivatives of lysergic acid and isolysergic acid compounds

Substance	Empirical formula (Mol. wt.)	m.p. (dec.)	$[\alpha]_D^{20}$	Typical crystallization and m.p. (dec.)
Lumi-lysergic acid-(I)		268°	+38° (in 0.5 N HCl) -13° (Pyridine)[203]	From water, short prisms
Lumi-lysergic acid-(II)	$C_{16}H_{18}O_3N_2$ (286.3)	190°	+11° (in 0.5 N HCl)	From water, rectangular plates
Lumi-isolysergic acid-(I)		217°	+47° (in 0.5 N HCl) +26° (Pyridine)[203]	From water, long thin prisms
Lumi-isolysergic acid-(II)		226°	-13° (in 0.5 N HCl) -21° (Pyridine)[203]	From water, rectangular plates
Lumi-lysergic acid-(I) amide		250°	+21° (in 0.5 N HCl)	From methanol, rods
Lumi-lysergic acid-(II) amide	$C_{16}H_{19}O_2N_3$ (285.3)	155°	—	From methanol, prisms
Lumi-isolysergic acid-(II) amide		154°	+22° (in 0.5 N HCl)	From methanol, thick pillars
Lumi-lysergic acid-(I) hydrazide	$C_{16}H_{20}O_2N_4$ (300.2)	241°	+2° (Pyridine)	From methanol, prisms[261]
Lumi-isolysergic acid-(I) hydrazide		244°	+54° (Pyridine)	From methanol polyhedra[261]
Lumi-lysergic acid-(I) diethylamide	$C_{20}H_{27}O_2N_3$ (341.4)	223°	-29° (Pyridine)	From benzene, thin prisms
Lumi-lysergic acid-(II) diethylamide		256°	-75° (Pyridine)	From chloroform/methanol, long prisms

177

The pharmacodynamic properties of the lumi compounds are compared to the parent substances less qualitatively altered than quantitatively attenuated[203].

Table 42
Lumi-derivatives of natural ergot alkaloids

Substance	Empirical formula (Mol. wt.)	m.p. (dec.)	$[\alpha]_D^{20}$ (pyridine)	Typical crystallization and m.p. (dec.)
Lumi-ergotamine-(I)	$C_{33}H_{37}O_6N_5$ (599.7)	247°	+14°	From methanol, pointed prisms
Lumi-ergotamine-(II)		192°	+2°	From acetone, polyhedra
Lumi-ergotaminine-(I)		217°	+68°	From chloroform/methanol, polyhedra
Lumi-ergotaminine-(II)		228°	−12°	From methanol/water, hexagonal plates
Lumi-ergosine-(I)	$C_{33}H_{37}O_6N_5$ (565.6)	205°	+30°	From ethyl acetate, needle tufts[261]
Lumi-ergocristine-(I)	$C_{35}H_{41}O_6N_5$ (627.7)	195°	+18°	From benzene, pointed prisms[261]
Lumi-ergocristine-(II)		171-174°	0° (±3°)	From methanol, short rods
Lumi-ergocristinine-(I)		234-236°	−116°	From methanol, needles
Lumi-ergcryptine-(I)	$C_{32}H_{43}O_6N_5$ (593.6)	201°	+39°	From ethyl acetate, polydedra[261]
Lumi-ergcornine-(I)	$C_{31}H_{41}O_6N_5$ (579.6)	202°	+32°	From ethyl acetate polydedra[261]
Lumi-ergobasine-(I)	$C_{19}H_{25}O_3N_3$ (343.4)	161°	−28°	From methanol/water, obliquely cut prisms
Lumi-ergobasine-(II)		128-138°	−19°	From methanol, needles
Lumi-ergobasinine-(I)		193-197°	+24°	From methanol, needles
Lumi-ergobasinine-(II)		125-135°	+18°	amorphous

d) Saturation of the double bond in the 2,3-position

The total synthesis of lysergic acid[96] led via the racemic 2,3-dihydro-lysergic acid. Later, the racemic 2,3-dihydrosetoclavine was also obtained totally synthetically[267]. In a simple manner, 2,3-dihydro derivatives have been prepared by partial synthesis by selectively hydrogenating lysergic acid or 9,10-dihydrolysergic acid compounds in the 2,3-position with zinc dust and concentrated hydrochloric acid[268]. This is one of the methods by which simple indoles have been reduced to the corresponding indolines[269, 270].

To carry out the reaction, concentrated hydrochloric acid is dropped slowly at room temperature and with vigorous stirring to an aqueous solution of a salt of the lysergic acid compound in question, after an excess of zinc dust had been added. The operation is continued until in a sample of the reaction solution the Keller's color reaction has become negative, indicating the transition of the indole to the indoline system.

Table 43

2,3-Dihydro derivatives of the lysergic acid and dihydrolysergic acid series

Compound	Empirical formula (Mol. wt.)	m.p. (dec.)	$[\alpha]_D^{20}$
2,3-Dihydro-d-lysergic acid	$C_{16}H_{18}O_2N_2$ (270.4)	—	—
2,3,9,10-Tetrahydro-d-lysergic acid	$C_{16}H_{20}O_2N_2$ (272.4)	298-300°	+ 86.5° (Acetic acid)
2,3-Dihydro-d-lysergic acid amide	$C_{16}H_{19}ON_3$ (269.5)	190-194°	+ 11° (CHCl₃)
2,3,9,10-Tetrahydro-d-lysergic acid amide	$C_{16}H_{21}ON_3$ (271.5)	239-241°	− 90° (CHCl₃)
2,3-Dihydro-d-lysergic acid diethylamide	$C_{20}H_{27}ON_3$ (325.5)	133-135°	+ 22° (Ethanol)
2,3-Dihydroergobasine	$C_{19}H_{23}O_2N_3$ (327.4)	193-194°	− 23° (CHCl₃)
2,3,9,10-Tetrahydroergobasine	$C_{19}H_{27}O_2N_3$ (329.4)	243-245°	− 169° (Pyridine)
2,3-Dihydro-d-lysergic acid (+)-butanolamide-(2)	$C_{20}H_{37}O_2N_3$ (341.5)	182-184°	− 19.5° (CHCl₃)
2,3-Dihydro-d-isolysergic acid (+)-butanolamide-(2)		152-153°	+ 263° (CHCl₃)
2,3-Dihydro-d-lysergic acid-cyclopentylamide	$C_{21}H_{27}ON_3$ (337.5)	212-214°	+ 7° (CHCl₃)
1-Methyl-2,3-dihydro-d-lysergic acid amide	$C_{17}H_{21}ON_3$ (283.4)	191-192°	− 76° (CHCl₃)
1-Methyl-2,3-dihydro-d-isolysergic acid amide		205-206°	+ 257° (CHCl₃)
1-Methyl-2,3-dihydro-d-lysergic acid diethylamide	$C_{21}H^{29}ON_3$ (339.5)	amorphous	—
1-Methyl-2,3-dihydroergobasine	$C_{20}H_{27}O_2N_3$ (341.5)	190-191°	− 79° (CHCl₃)
1-Methyl-2,3,9,10-tetrahydro-ergobasine	$C_{20}H_{29}O_2N_3$ (343.5)	Bimaleinate: 165-170°	—
1-Methyl-2,3-dihydro-d-lysergic acid (+)-butanolamide-(2)	$C_{21}H_{29}O_2N_3$ (355.5)	200-202°	− 95° (CHCl₃)
1-Methyl-2,3,9,10-tetrahydro-d-lysergic acid-(+)-butanolamide-(2)	$C_{21}H_{31}O_2N_3$ (357.5)	215-218°	− 102° (Dioxane)

Upon hydrogenation of the double bond in the 2,3-position, a new center of asymmetry is formed at C3. The reduction with Zn/HCl is stereospecific; only one of the two theoretically possible epimeric dihydro

179

derivatives is formed. On the basis of reaction-mechanistic considerations, the hydrogen atom in position 3 can very probably be attributed ß-perpendicular, axial position, as expressed in Formula Scheme 29.

Formula Scheme 29

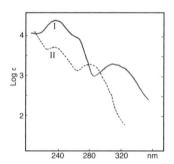

Spatial formula of the 2,3-dihydro-*dl*-lysergic acid derivatives

The reduction can also be carried out on 1-alkylated lysergic acid derivatives. The 2,3-dihydrolysergic acid compounds are dibasic. The NH group in the 1-position can be easily acylated.

Table 43 compiles a number of 2,3-dihydrolysergic acid derivatives[271].

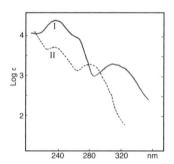

Figure 20. UV spectra in methanol
I = 2,3-dihydro-d-lysergic acid amide, II = 2,3,9,10-tetrahydro-d-lysergic acid amide

The UV spectra of the 2,3-dihydro derivatives are characteristically different from those of the corresponding lysergic acid and 9,10-dihydrolysergic acid compounds (cf. Figure 14). Figure 20 shows the UV spectra of the 2,3-dihydrolysergic acid and the 2,3,9,10-tetrahydrolysergic acid derivatives[203].

Formula Scheme 30

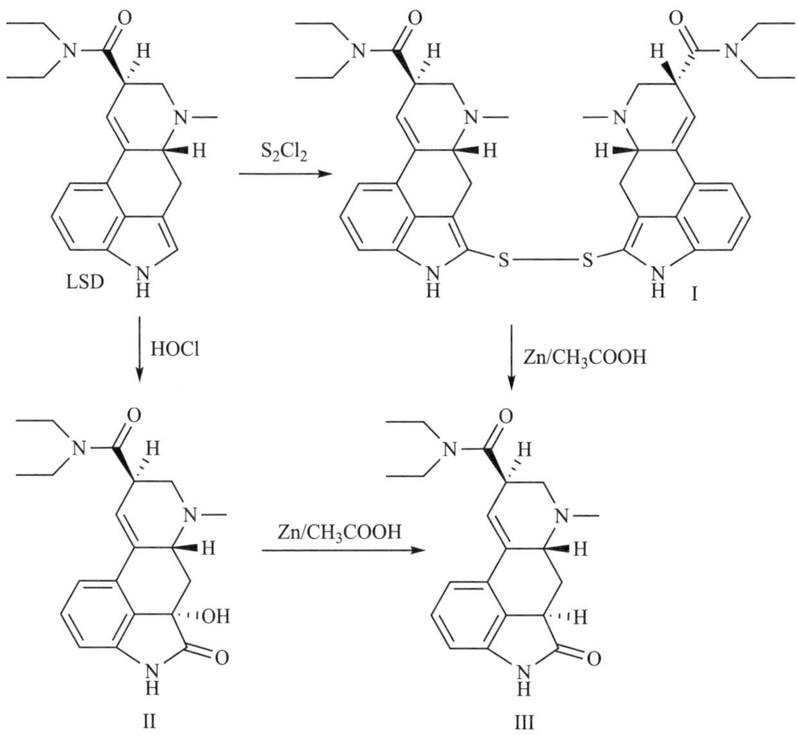

The 2,3-dihydrolysergic acid derivatives can be dehydrogenated with deactivated Raney nickel with the addition of sodium arsenate[96] or with mercury(II) acetate[268] to give the lysergic acid derivatives used as starting materials. Dehydrogenation with mercury(II) acetate proceeds only with moderate yields for the derivatives unsubstituted on indoline nitrogen, but with good yields for the 1-alkylated compounds.

e) Oxidation in the 2,3-position

In connection with the identification of biological transformation products of lysergic acid derivatives (cf. next Section VIII, 2), oxidation on the pyrrole ring of the indole system was investigated. 2-Oxo-2,3-dihydrolysergic acid diethylamide (III), obtained by the action of a liver microsome preparation on LSD, could be prepared chemically by two different methods, according to Formula Scheme 30 (above).

According to the procedure leading via a disulfide compound (I), $C_{40}H_{48}N_6O_2S_2$, m.p. 182–184°, $[\alpha]_D^{20}$ = -1020° (pyridine), a preparation could be acquired that, like the biological oxidation product, was obtained only in the amorphous state[272]. Oxidation of LSD with hypochlorite[273] first gives the dioxindole compound 2-oxo-3-hydroxy-2,3-dihydro-*d*-lysergic acid diethylamide (II), which can be reduced with zinc dust in dilute acetic acid to 2-oxo-2,3-dihydro-*d*-lysergic acid diethylamide (III) and could, by this method, be obtained in crystallized state[273]. From benzene colorless needles, m.p. 235–237° (dec.), $C_{20}H_{25}O_2N_3$ (339.4), $[\alpha]_D^{20}$ = + 10° (pyridine). UV spectrum in methanol: maximum at 256 mμ (log ε = 4.06). However, the yield, especially in the last step, is very low, because naphthostyril and indole derivatives are also formed by the zinc dust reduction.

The dioxindole derivatives of the compounds of the lysergic acid and dihydrolysergic acid series compiled in Table 44 were also prepared by oxidation with hypochlorite[273].

Table 44
Dioxindole derivatives of the lysergic acid and dihydrolysergic acid series

Compound	Empirical formula (Mol. wt.)	m.p. (dec.)	$[\alpha]_D^{20}$ (Pyridine)
d-Lysergic acid diethylamide	$C_{20}H_{25}O_3N_3$ (355.4)	155-160°	− 50°
1-Methyl-*d*-lysergic acid diethylamide	$C_{21}H_{37}O_3N_3$ (269.5)	230-231°	− 50°
Ergotamine	$C_{33}H_{35}O_7N_5$ (613.7)	amorphous	+ 15°
Dihydro-*d*-lysergic acid-(I) methyl ester	$C_{17}H_{20}O_4N_2$ (316.4)	252°	− 72°
Dihydroergotamine	$C_{33}H_{37}O_7N_5$ (615.7)	236°	− 40°

No definite statement can yet be made about the steric position of the hydroxyl group at C atom 3. Addition of hypochlorous acid to endocyclic double bonds normally leads to an axial position of the OH group. Observations on the model show that an axial OH group is located in the ß-position in the pseudo-armchair form of the ring C, but in the α-position in the pseudo-boat form. Since the former form will be energetically favored, the ß-position of the OH group may be assumed to be probable in the dioxindole compounds.

VIII. Biological transformation of ergot alkaloids

In the context of research on the conversion of pharmacologically less effective clavine alkaloids into derivatives of the lysergic acid series, microbiological transformation of ergot alkaloids has also been investigated. Furthermore, the great importance of certain ergot alkaloids and some of their derivatives in pharmacology and therapy has aroused interest in the metabolism of these substances in the animal organism. In this direction, studies on isolated organ preparations and on whole animals in vivo are available.

1. Microbiological hydroxylation

The Mexican psychoactive fungus *Psilocybe semperviva* Heim et Cailleux hydroxylates tryptophan in the 4-position[275] during the biosynthesis of psilocybin, a hallucinogenic agent of this magic drug[274]. When this basidiomycete acted on the ergot alkaloids of the clavine series, which are foreign to it and in which the 4-position of the indole system is already occupied, hydroxylation occurred in the 8-position of the ergoline system[276] (cf. Formula Scheme 31). Elymoclavine added to shaking cultures or standing cultures of the fungus *Psilocybe semperviva* Heim et Cailleux was converted into penniclavine and isopenniclavine. In parallel with this hydroxylation, degradation proceeds to products that no longer show color reactions of the ergot alkaloids. From agroclavine, setoclavine was formed in addition to very little isosetoclavine (Tables 45–47). These microbiological oxidations thus proceed formally analogous to the corresponding chemical transformations reported by S. Yamatodani[180] and A. Hofmann et al.[18]. In both processes, the molecules appear to react with OH[+] in the condition indicated in the formula diagrams below (Formula Scheme 31), the state that must also be responsible for the known rearrangement reactions (e.g., elymoclavine to lysergol) in alkaline medium.

Table 45
Hydroxylation of elymoclavine with *Psilocybe semperviva* in shaking culture

Days after addition of the elymoclavine	Total alkaloid content (colorimetric)	Composition		
		Elymoclavine %	Penniclavine %	Isopenniclavine %
0	100	100	0	0
2	48	80	12	8
4	27	50	30	20
7	10	20	50	30
10	4	20	50	30

In hydroxylation experiments with ergotamine and ergobasine as substrate, no defined conversion products could be captured.

Table 46
Hydroxylation of elymoclavine with *Psilocybe semperviva* in standing culture

Days after addition of the elymoclavine	Total alkaloid content (colorimetric)	Composition		
		Elymoclavine %	Penniclavine %	Isopenniclavine %
0	100	100	0	0
4	83	97	2	1
7	80	95	3	2
14	62	85	10	5

Table 47
Hydroxylation of agroclavine with *Psilocybe semperviva* in shaking culture

Days after addition of the agroclavine	Total alkaloid content (colorimetric)	Composition		
		Agroclavine %	Setoclavine %	Isosetoclavine %
0	100	100	0	0
2	15	97	2	1
4	2	95	3	2
7	0	85	10	5
New Attempt 1	60	25	75	Trace
Control (uninoculated) 1	90	100	0	0

Formula Scheme 31

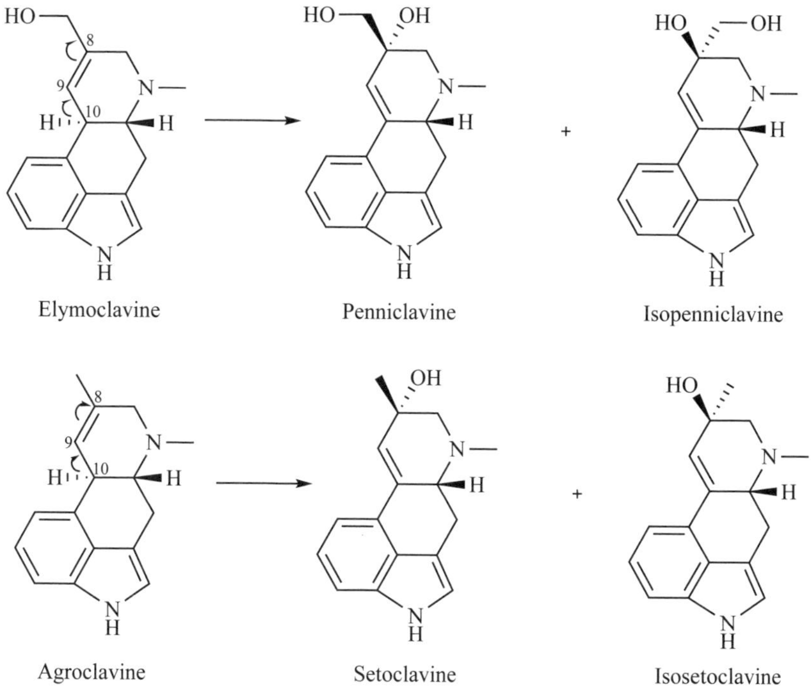
HO
8
9
N
10
H
H
N
H
Elymoclavine
HO
OH
N
H
N
H
Penniclavine
+
HO
OH
N
H
N
H
Isopenniclavine
8
9
N
10
H
H
N
H
Agroclavine
OH
N
H
N
H
Setoclavine
+
HO
N
H
N
H
Isosetoclavine

2. Experiments with organ preparations

The biochemical transformations in the animal organism and by the enzyme systems of isolated organs have been studied most thoroughly in the case of using *d*-lysergic acid diethylamide. The extraordinarily high and specific psychotomimetic activity of this substance (cf. Section D) made the questions about the metabolism of this ergot derivative particularly important, because it could be related to the biochemistry of psychic functions.

Comparative studies with organ sections of the liver, brain, kidney, spleen, and muscles of the guinea pig showed that only the tissue of the liver is capable of metabolizing LSD[208]. In the liver, it was the microsomes, together with soluble factors, that caused the chemical transformation. LSD is converted by an enzyme system present in the liver microsomes of the guinea pig into a substance that could be identified as 2-oxo-2,3-dihydro-d-lysergic acid diethylamide (cf. Formula Scheme 30, III)[277]. The structure of this biological oxidation product was confirmed by comparison with synthetic III (cf. Section VII, 3e). Oxidation in the 2-position inactivates LSD. 2-Oxo-2,3-dihydro-d-lysergic acid diethylamide is pharmacologically ineffective and has also lost the hallucinogenic property of the parent compound.

3. Transformations in the animal organism

The studies on the distribution and metabolism of ergot alkaloids and their derivatives in the organism have been undertaken to gain insight into the mechanism of action associated with the pharmacology of these substances. In most of the work, the presence or the disappearance of the alkaloid in the body fluids and in the various organs have been determined only by means of the pharmacological action of the substances concerned, as with the help of adrenolysis in the guinea pig seminal vesicle of the genuine and hydrogenated peptide alkaloids[278] or serotonin antagonism on the isolated rat uterus in the case of lysergic acid diethylamide[279]. These results are discussed in the context of pharmacology in Section D, as are the studies with radiolabeled lysergic acid diethylamide[280–282], in which no defined transformation products were detected either.

So far, only *M. B. Slaytor* and *S. E. Wright*[283] have succeeded in elucidating the chemical nature of metabolites. They were able to separate the conversion products of ergobasine and lysergic acid diethylamide from rat bile by paper chromatography and determine their chemical structure.

LSD was administered intravenously at doses of 3 mg/kg and ergobasine at doses of 3 mg/kg and 45 mg/kg to rats under urethane anesthesia, and bile was collected thereon through a cannula for six hours.

From the bile of LSD-treated animals, two metabolites could be separated by paper chromatography and small amounts could be obtained in crystallized form from the paper chromatograms. The highly polar conversion products could be rearranged into each other and cleaved by ß-glucuronidase into less polar compounds for which the structures Ia and IIa of 12-hydroxy-*d*-lysergic acid diethylamide and 12-hydroxy-*d*-isolysergic acid diethylamide could be deduced on the basis of spectral properties and color reactions (cf. Formula Scheme 32). The metabolites are therefore probably the glucuronides of 12-position hydroxylated LSD (Ib) and iso-LSD (IIb).

Ergobasine was metabolized in a corresponding manner when administered at the low dose. From the physical and chemical properties of the two metabolites, the structures Id and IId could be inferred. The 12-hydroxy-ergobasine (Ic) formed from glucuronide Id on exposure to ß-glucuronidase was identified by comparison with a synthetically prepared compound.

At the higher dosage of ergobasine, in addition to these two metabolites, eight further spots were found on the paper chromatogram, two of which most probably belong to ergobasine and ergobasinine and two to the glucuronic acid glycosides of these alkaloids. It can be assumed that the glucuronic acid residue in this case is located at the hydroxyl of the propanolamine side chain.

Formula Scheme 32

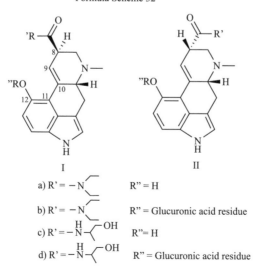

I   II

a) R' = −N⟨   R" = H

b) R' = −N⟨   R" = Glucuronic acid residue

c) R' = −N(H)⟨−OH   R" = H

d) R' = −N(H)⟨−OH   R" = Glucuronic acid residue

The metabolization of ergobasine and LSD in the rat by hydroxylation at the 12-position is understandable from a chemical point of view, since the 12-position is considered a point of high electron density in the indole system[284].

D. THE PHARMACOLOGY AND THERAPEUTIC USE OF THE ERGOT ALKALOIDS AND THEIR DERIVATIVES

There are probably only a few natural substances that are characterized by such a broad spectrum of pharmaceutical activity as ergot alkaloids. They are used therapeutically in obstetrics and gynecology, in internal medicine, and in psychiatry.

The first medicinal use of ergot as an oxytocic dates back to the Middle Ages, as already mentioned in the introductory chapter on the history of ergot. This indication, which found its way into orthodox medicine after the well-known publication of the American physician *John Stearns* in 1808, entitled "Account of the Pulvis parturiens, a Remedy for Quickening Childbirth,"[48] was soon dropped again upon the recommendation of *D. Hosack*[49] because, as we know today, due to the variable active ingredient content of the drug, overdoses frequently occurred, which led to uterine convulsions and the death of the child. From then on, stopping postpartum hemorrhage remained the main indication for ergot preparations in obstetrics. The first chemically uniform ergot alkaloid, ergotamine, from *A. Stoll*[58–60] was also introduced into therapy under the brand name "Gynergen"®, initially exclusively for this indication. Of the very numerous publications in this field, only the first pharmacological and clinical introductory works by *K. Spiro*[62, 350] and by *H. Guggisberg*[303, 304] can be cited here. However, in the basic pharmacodynamic studies by *E. Rothlin* on ergotamine[305–308, 355], *E. Rothlin* and *A. Cerletti*[309] showed that this alkaloid has other therapeutically valuable properties. The specific inhibitory effects on sympathetic functions of the autonomic nervous system, which *H. H. Dale* had discovered in ergotoxine[55], are also attributed to ergotamine. With these investigations, the pharmacology of ergot alkaloids began to expand. This was followed by a great deal of work in many laboratories, which revealed the entire breadth of the pharmacological spectrum of action of ergot alkaloids. To facilitate access to the original studies, which are too numerous to be cited individually here, reference is made here to later summary publications[278, 310, 311].

The pharmacological action of all ergot alkaloids can be ascribed, according to *A. Cerletti*[312], mainly to the six effect categories schematically shown in Figure 21, representing a combination of these individual components. According to the site of their action, which varies from case to case, three groups of effect components can be distinguished: peripheral-muscular, neurohumoral and central-nervous.

Figure 21. Pharmacological action of ergot alkaloids

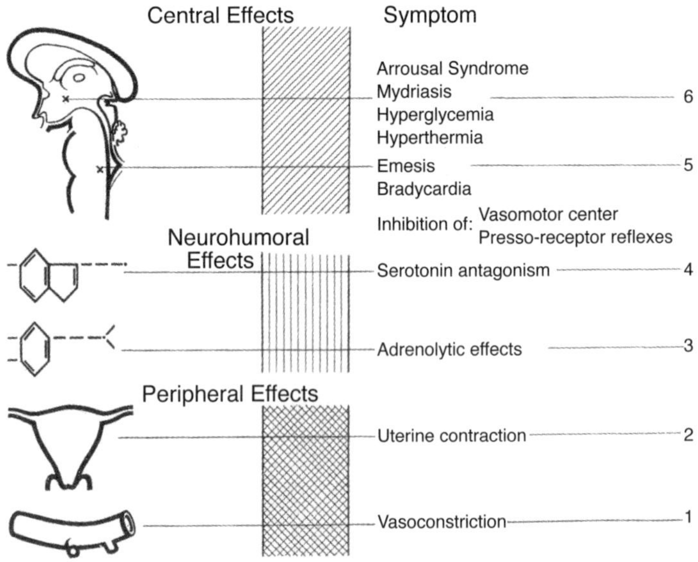

The peripheral smooth-muscle effect manifests itself in vasoconstriction and uterine contraction. The classical indication of ergot alkaloids, their use in obstetrics as hemostatic and labor-promoting drugs, is based on the uterus-contracting effect.

Neurohumoral effects the adrenalin antagonism and the serotonin antagonism[392]. The adrenaline-antagonistic, adrenolytic effect determines the application of ergot preparations as highly effective specific sympatholytics in internal medicine. The serotonin-antagonistic effect could be selectively elucidated in certain ergot derivatives, as will be detailed below.

The central effects can be localized at two separate sites: the medulla oblongata and the diencephalon.

In the medulla oblongata, the vasomotor center is affected in an ettenuating sense. This is related to vasodilatory and antihypertensive, as well as bradycardic effects of certain ergot preparations. Certain ergot derivatives stimulate the emetic center in the medulla oblongata.

In the diencephalon, especially in the hypothalamus, sympathetic pathways are stimulated, which leads to a comprehensive excitation syndrome characterized by mydriasis, hyperglycemia, hyperthermia, tachycardia, etc. This syndrome is related to the psychotomimetic effects of certain ergot derivatives.

This pluripotency inherent in the natural ergot alkaloids is the reason why chemical interventions in the delicate structure of these substances do not only result in a weakening, or a loss of efficacy (per se), but also in a change of effect. This change in effect is due to the fact that, in the case of virtually complete loss of one or more partial effects as a result of chemical modification, other partial effects remain or are even intensified. The chemical modification products therefore usually turn out to be less comprehensively effective than the parent substances, i.e., the natural parent alkaloids; conversely, however, they are more selectively effective with respect to the remaining active components.

To illustrate the effects of the various types of ergot alkaloids and their derivatives, the graphical representation, as used by *A. Cerletti*[312] taking into account the six main effects discussed, is very suitable. For the construction of the curves of these effect spectra, a relative scale was chosen for each compound, starting with the smallest effective dose (top) and ending with the 100% lethal dose (bottom). The maxima thus indicate which of the six main effects listed are particularly prominent in a particular compound and thus dominate the overall effect picture.

Ergotamine and the other peptide-type alkaloids

Figure 22 shows the action spectrum of ergotamine. From this it is evident that in ergotamine the proportional contributions of the various components of action are largely balanced. This alkaloid thus possesses the full effect of ergot, in that both the uterine contractile effect and the attenuation of the adrenergic system and the central effects produced by inhibition of the vasomotor center are present[278, 309]. Only the stimulation of the central sympathetic structures, the central excitation syndrome, is less pronounced.

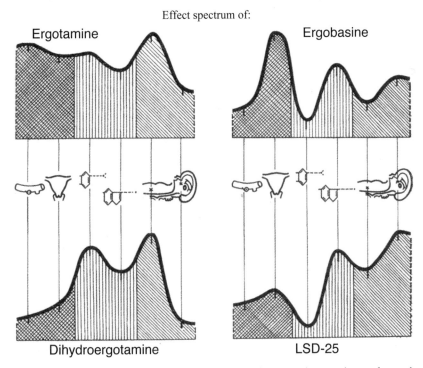

Figure 22: For the construction of the curves, a relative scale was chosen for each compound, starting with the smallest effective dose (top), and ending with the 100% lethal dose (bottom). The maxima thus indicate which of the six main effects listed are particularly prominent and thus dominate the overall effect picture. For the explanation of the symbols between the curves see Figure 21.

Ergotamine acts on the uterus in vivo and in vitro[62, 278, 313]. Its action is characterized by an increase in tone and an amplification of rhythmic contractions that occur after a shorter or longer latency period, depending on the route of administration, and last for several hours. The sensitivity of the uterus to ergotamine and the other ergot alkaloids varies greatly with its status. The gravid uterus is much more sensitive, so that even very small doses of alkaloid, given shortly before birth or immediately post partum, are specifically effective without side effects. Since contractions can also be induced in isolated strips of uterine musculature, this is a directly muscular effect.

The effect on smooth muscle is also manifested on the vascular apparatus, in that ergotamine produces an increase in blood pressure both in the whole animal and in the decapitated or spinalized cat. Thus, since the effect occurs even after the vasomotor center is cut off, this is also a direct muscular effect that manifests itself in a contraction of the small blood vessels. With ergotamine, this peripheral vasoconstrictor effect always masks the centrally induced reduction in vascular tone.

The sympatholytic effect of ergotamine is expressed in the fact that the responsiveness of the organs to sympathetic and adrenergic stimuli is inhibited by ergotamine. This sympatholytic effect extends to both sympathetically promoted and sympathetically inhibited functions[278]. This sympatholytic effect can be seen particularly clearly, for example, in the rabbit uterus or in the isolated seminal vesicle of the guinea pig, which is contracted by adrenaline. Smallest amounts ergotamine cause a complete abolition of this adrenaline effect. The isolated seminal vesicle of the guinea pig serves as a biological test object for the quantitative differentiation of the sympatholytic action of the various ergot alkaloids and their derivatives[351]. The abolition by ergotamine of the tonic reduction induced by adrenaline at the intestine is an example of antagonism to a sympathetically inhibited function.

The hyperglycemia and basal metabolic rate increase induced by adrenaline is reversed by ergotamine and the other peptide alkaloids[310].

The slowing of the cardiac rhythm, the inhibition of the carotid sinus reflex, as well as the emetic property are due to central effects of ergotamine in the medulla oblongata. Another central effect of ergotamine that should be emphasized is mild sedation, which is expressed primarily in a potentiation of sedative and sleep-inducing drugs[310].

This action pattern justifies the use of ergotamine in obstetrics as a hemostatic agent and in internal medicine and neurology as a vegetative depressant drug. The widespread use of ergotamine to combat migraine is most likely due to its pronounced vasoconstrictor property[352]. Ergotamine is most commonly prescribed in the form of its tartrate in tablets of 0.5 mg or 1.0 mg and as a solution for injection, 1 cm^3 containing 0.5 mg ergotamine tartrate (Gynergen®). In obstetrics, the combination with the water-soluble ergobasine, which is also very rapidly effective perorally, is advantageous (0.25 mg ergotamine tartrate + 0.125 mg ergobasine tartrate in 1 cm^3 [Neo-Gynergen®]). The most extensive use of ergotamine is in neurology and psychiatry and also in internal medicine for attenuation of the autonomic nervous system in combination with the parasympathetically active belladonna alkaloids and centrally sedating substances such as barbiturates, e.g., in the form of the special preparation Bellergal®. In migraine therapy, especially to reduce the pain during migraine attacks, the combination of ergotamine with caffeine has proven effective, e.g., as Cafergot®.

The other peptide-type alkaloids, ergosine, and the alkaloids of the ergotoxine group, show a spectrum of action similar to that of ergotamine. The central effects, especially those based on stimulation of sympathetic centers in the midbrain, tend to be somewhat enhanced compared with the other components of action, resulting in a higher acute toxicity compared with ergotamine.

Some study results are available on the absorption and excretion of ergotamine and the other peptide-type alkaloids, but nothing is yet known about their biochemical transformation in the organism.

Metabolic products are only known from ergobasine and lysergic acid diethylamide (see Section C, VIII/3).

Based on the color reaction with *p*-dimethylaminobenzaldehyde, the rapid disappearance of intracardially applied ergotamine, ergotoxine, and ergonovine (ergobasine) from the blood and muscle tissue of guinea pigs was established[314]. Using the very sensitive biological detection method with the use of adrenolytic activity on the isolated guinea pig seminal vesicle, only 0.001% of the applied alkaloid could be detected in the urine after intravenous injection of ergocornine tartrate in rats[315].

Using the same detection method, the disappearance of intravenously applied dihydroergocryptine from the blood of rabbits was followed[278]. After only two minutes, only 2.2% of the alkaloid was still present in the blood. However, relatively large amounts of the alkaloid could be detected in the liver, kidney, spleen, and muscle tissue. The active substance was also found in the cerebrospinal fluid, but not in the brain.

Conclusions on the degree of absorbability can be drawn from clinical observations. The peptide-type alkaloids are only moderately absorbed in the gastrointestinal tract; the effective oral dose is 8–10 times greater than that administered parenterally. In contrast, the low-molecular-weight ergot alkaloids, e.g., ergobasine, are well absorbed even when administered perorally, and the effect is prompt.

Ergobasine and *d*-lysergic acid (+)-butanolamide-(2)

The action spectrum of ergobasine (see Figure 22) is very different from that of the peptide alkaloids, from which this alkaloid also differs chemically quite substantially. With hardly any adrenolytic activity and greatly attenuated central effects, but remarkable anti-serotonin action, ergobasine's uterine, hemostatic, and oxytocic effects are quite prominent. Ergobasine is therefore used almost exclusively in obstetrics. A limited use and some success as a migraine remedy is probably related to the vaso-contracting action of this alkaloid, which is, however, only moderately pronounced. Ergobasine is most commonly prescribed in the form of the acidic maleate (Ergometrine Maleate [BP]; Ergonovine Maleate [USP]) in tablets of 0.2 or 0.5 mg and in ampoules of 0.2 or 0.5 mg in 1 cm^3.

When the 2-aminopropanol residue in the natural ergot alkaloid ergobasine was replaced by the 2-aminobutanol residue[157], a partially synthetic alkaloid was obtained which has a virtually identical action pattern to ergobasine. The smooth muscle effects, the effects on the uterus, are predominant. Compared with the natural ergot alkaloid, the synthetic d-lysergic acid (+)-butanolamide-(2) exhibits a rather even more intense and longer-lasting uterine effect. It is widely used in obstetrics in the form of the special preparation "Methergine"®.

Table 48

Overview of the pharmacological properties of natural ergot alkaloids and their dihydro derivatives

Alkaloid	Direct action on smooth muscle		Central vasodilator effect	Adreno-sympathic-olytic action	Bradycardic effect	Mouse toxicity (i.v.)	
	uterotonic	vasocon-strictive				LD50 mg/kg	Relative toxicity (Ergotamine = 100%)
Ergobasine	+++	+	0	0	0	126	56%
Methylergobasine (Lysergic acid (+) butanolamide-(2))	++++	(+)	0	0	0	80	87%
Ergotamine	+++	+++	(+)	++	++	70	100%
Ergocornine Ergocristine Ergocryptine $\}$ 1)	+	+++	+++	++	++	40	175%
Dihydro-ergotamine	+	++ or, depending on the functional status of the assay +	+ or, depending on the functional status of the assay ++	+++	++	118	59%
Dihydro-ergocornine Dihydro-ergocristine Dihydro-ergocryptine $\}$ 2)	(+)	0 or (+)	++++	++++	++	174	40%

1) Mixture of the 3 alkaloids = ergotoxine
2) Mixture of the 3 alkaloids (as methanesulfonate salts) = Hydergine

9,10-Dihydro derivatives of peptide alkaloids

Due to the saturation of the carbon double bond in the 9,10-position in the lysergic acid part of the alkaloids of the ergotamine and ergotoxine group[120], the pharmacological properties undergo fundamental changes, which were mainly investigated by *E. Rothlin* and coworkers[310, 353, 354]. Figure 22 shows the spectrum of action of dihydroergotamine. The smooth muscle effects, vasoconstriction, and uterine effects, i.e., the classical ergot effects, as well as the central sympathetic stimulation, are so strongly weakened that they are practically no longer apparent. Instead, a pronounced sympatholytic-adrenolytic effect and the attenuation of the vasomotor center now determine the effect pattern, which is characterized by vasodilatation, blood pressure reduction[316], and a certain sedation.

The dihydro derivatives of the peptide alkaloids of the ergotoxine group show similar action spectra as dihydroergotamine. Adrenolysis, vasodilatation, and blood pressure reduction are even more pronounced. In Table 48, the most important pharmacological properties of the dihydro derivatives are compared with those of the corresponding genuine ergot alkaloids in their potency. (Ergobasine and its next higher homologue, *d*-lysergic acid (+)-butanolamide-(2) [= methylergobasine], are also included in the comparison.) Furthermore, the table contains numerical values on toxicity in mice. It is evident from this that hydrogenation not only causes a fundamental shift in the strength of the individual active components, but that toxicity also undergoes a considerable attenuation.

Due to these pharmacological properties, the dihydro derivatives of the peptide alkaloids find versatile clinical application alone or in combination with other drugs. Dihydroergotamine, for example, in the form of methanesulfonic acid salt, is the active ingredient of the pharmaceutical specialty "Dihydergot"® and is used in the treatment of migraine, neurocircular dystonia, dysmenorrhea, for the treatment of disorders associated with herpes zoster and other special indications. The dihydro derivatives of ergocristine, ergocryptine and ergocornine, due to their pronounced vasodilatory effects, find widespread therapeutic application for peripheral circulatory disorders of various etiologies, cerebral circulatory disorders[317], and in certain forms of hypertension. They have proved particularly useful in geriatrics. The most widely used special preparation on this basis is "Hydergine"®, which consists of equal parts of these three dihydro alkaloids in the form of their methanesulfonates. Hydergine, thanks to its adrenolytic, vasodilator, antipyretic, and metabolic stimulant properties, is also used in drug combinations to induce artificial hibernation.

d-Lysergic acid diethylamide (LSD-25)

One compound that again has quite different effects than the natural ergot alkaloids or their dihydro derivatives is the semisynthetic d-lysergic acid diethylamide, which has become known under the experimental name LSD-25 (Delysid®).

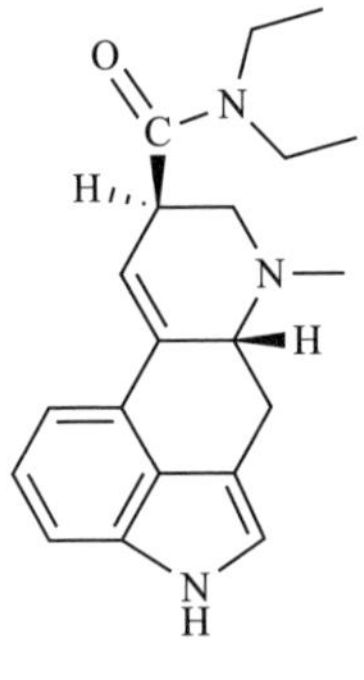

LSD-25

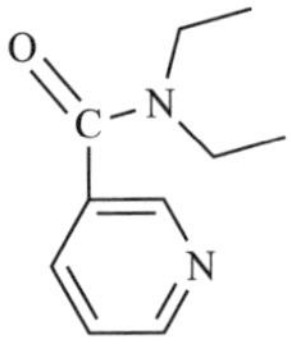

Coramine

This substance was first prepared in 1938 in the pharmaceutical-chemical laboratories Sandoz and published together with a large number of simple acid amide-like derivatives of lysergic acid in 1943[157]. The lysergic acid diethylamide was synthesized at that time in the hope of obtaining an analeptic, which could be expected because of the structural relationship of ring D of lysergic acid with the known analeptic "Coramine"®, the diethylamide of nicotinic acid. During the renewed study of this substance, A. Hofmann discovered in 1943 its extraordinarily high and specific efficacy on the human psyche, which he was able to confirm in a subsequent planned self-experiment with 0.25 mg d-lysergic acid diethylamide tartrate[318, 319]. The first basic psychiatric study on the clinical features of LSD intoxication in healthy and mentally ill people was carried out by W. A. *Stoll* in the psychiatric university hospital in Zürich[320]. The great interest that this substance has found in pharmacology, experimental psychiatry, and, more recently, increasingly in psychotherapy, is evident from the fact that already more than 1,500 scientific publications have appeared on this subject.

LSD is the prototype of that special subgroup of psychotropic drugs which have been called phantastica, hallucinogens, psychotomimetics, psychedelics, psychodisleptics, and so on. Representatives of this group of drugs have been known for a long time. They are the active principles of magic plants, of magic drugs, which have played a significant role in the cultural history of mankind since ancient times, and which are still used today by primitive peoples in religious ceremonies and for healing purposes. The three most important magic drugs, which were used already in the ancient Indian cultures of Central America, and which are still used today in remote areas of Mexico by the healing priests (shamans) of the Indians, may be mentioned here as examples: Peyotl (*Lophophora williamsii*, aka Peyote), a cactus whose main active ingredient is mescaline[321]; Teonanacatl (*Psilocybe* spp.), the sacred mushroom of the Aztecs, for whose hallucinogenic effects the two indole compounds psilocybin and psilocin are responsible[322]; and Ololiuqui, the seeds of morning glory plants from which d-lysergic acid amide (ergine) and d-isolysergic acid amide (isoergine) have been isolated as the main active ingredients along with some other ergot alkaloids[78]. The psychotropic effect of lysergic acid amide, which contains a markedly narcotic component, was observed long before the discovery of this substance as a natural alkaloid in "Ololiuqui"[323, 324].

The synthetic psychotomimetic lysergic acid diethylamide thus has close relatives in nature that are responsible for the psychotropic effect of a Mexican magic drug. The most effective psychotomimetics known to date, namely LSD, the ololiuqui active ingredients and the psilocybin and psilocin from teonanacatl, have common characteristic structural features (cf. Formula Scheme 33). They are tryptamine derivatives substituted in the 4-position.

Formula scheme 33

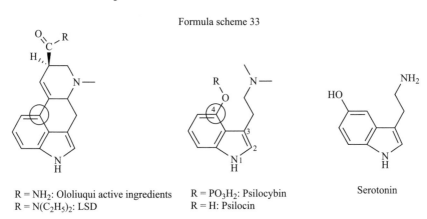

R = NH₂: Ololiuqui active ingredients
R = N(C₂H₅)₂: LSD

R = PO₃H₂: Psilocybin
R = H: Psilocin

Serotonin

The close structural relationship of these psychotomimetics with the neurohumoral factor serotonin (5-hydroxytryptamine), which is enriched in the brain, especially in the hypothalamus, and which plays a role in the chemistry of central nervous processes[325], indicates that certain indole structures are important for the biochemistry of mental processes[326].

The characteristic and fascinating thing about the psychotomimetics is their specific, very profound effect on the psyche, which these substances trigger without causing serious disturbances in the autonomic nervous system or in the physical functions. Both the experience of the environment, its forms and colors, and of one's own mental and physical personality are changed in a fantastic way by the psychotomimetics. But not only space, also time, the other basic factor of human existence, is experienced under the influence of these substances in a completely different way than in the normal state. Often the sense of time is completely suspended. At the same time, all these fundamental changes in the world view are experienced quite consciously. The test subject enters with full consciousness into other worlds, into a kind of dream world, which, however, is felt to be quite real, indeed usually even more real, more intense, and more meaningful than the ordinary everyday world. The senses, especially the sense of sight, are sensitized in an unusual way, and this stimulation can sometimes increase, especially at higher doses, to the point of triggering hallucinations and visions. It seems that the control mechanisms normally functioning in a limiting sense are suspended in our nervous system and that then, as a consequence, from outside the whole cosmic wealth of images and from within the subconscious, a tremendous stream of sensations and memories penetrates into the consciousness. It is not surprising that this experience in its vehemence, fullness, and strangeness was often interpreted in the sense of a religious revelation, which makes the use of these drugs for cultic purposes understandable.

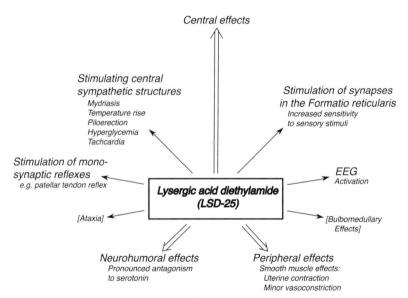

Figure 23. The pharmacological effect of LSD

The discovery of LSD, which produced such psychic changes already in doses of 20–50 millionths of a gram, i.e., was about 1,000–5,000 times more effective than the already known mescaline, aroused intensive research activity in the field of psychotomimetics. Along with the introduction of chlorpromazine, which could be used to attenuate psychic functions in a hitherto unfamiliar way, a new line of research emerged for which the term *psychopharmacology* was coined.

The pharmacological effects of LSD, which have been studied mainly in the pharmacologic laboratories of Sandoz AG, Basel, by *E. Rothlin*, *A. Cerletti*, and coworkers[327–329], is represented by the activity curve in Figure 22 and by the Scheme in Figure 23.

Although marked peripheral effects are also present, in that LSD still has a fairly strong uterine activity, the pattern of effects, except for the marked antagonism to serotonin, is entirely dominated by the stimulation of central sympathetic structures in the hypothalamus. These resulting central effects are manifold. They can be summarized under the term ergotropic stimulation syndrome. This stimulation syndrome includes:

a) Activation of the wave pattern in the EEG (electroencephalogram, i.e., in the curvilinear recording of the action potential activity of the brain);

b) Stimulation of the synapses in the reticular formation, resulting in an increase in sensitivity to sensory stimuli;

c) Stimulation of central sympathetic structures, manifested by mydriasis, increase in temperature, piloerection, etc; and

d) Stimulation of monosynaptic reflexes, e.g., the patellar tendon reflex.

Ataxia and bulbo-medullary effects, such as vomiting, occur only after very high, toxic doses.

However, LSD does not show the activation across the board. In certain tests and on certain animals, there are also markedly depressive effects. For example, the barbiturate narcosis of the mouse and the rat is intensified by LSD, and the body temperature and oxygen consumption are lowered.

In general, the doses necessary to achieve the described pharmacological effects in animals are much higher than the amounts that are psychologically effective in humans. Only the rabbit is an exception, in which certain vegetative effects, e.g., temperature increase, can be induced with the same minimum dosages, namely with 0.5–1 µg per kg body weight.

The toxicity of LSD in animals is very low compared to its psychic potency in humans. It varies greatly from species to species. The LD_{50} is 46 mg/kg in mice, 16 mg/kg in rats, and 0.3 mg/kg in rabbits. Death occurs by respiratory arrest[330]. This compares with the effective dose of 0.0005 mg/kg in humans. It is not straightforward to compare toxicity in animals and efficacy in humans. Nevertheless, from these figures it may be concluded that the specificity of the psychic effect is quite unique.

With such a specific psychotomimetic as LSD, of course, the distribution in the organism is of particular interest. One would expect to find such a compound enriched in the brain. However, the corresponding investigations showed that this is not the case.

Figure 24. Distribution and excretion of LSD-^{14}C.

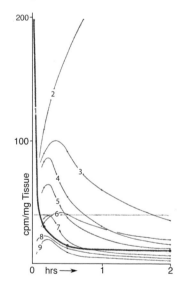

1 blood

2 small intestine

3 liver

4 kidney, adrenal gland

5 lung, spleen, pancreas

6 viscera;

7 heart

8 muscle, skin

9 brain

Distribution in the organism and excretion was determined in the mouse using ^{14}C-labeled LSD[280–282, 331]. As shown in Figure 24, LSD rapidly disappears from the blood[281] after i.v. injection and is then found in the various organs. Surprisingly, the concentration in the brain is the smallest. The concentration in the organs, which peaks after 10–15 minutes, decreases very rapidly. An exception is the small intestine. Here the activity increases to a maximum within two hours. The radioactive substance then moves on with the intestinal contents. Thus, most of the excretion, namely about 80%, happens via liver bile through the intestinal tract.

Extraction experiments carried out on various organs two hours after administration showed that only 1-10% of the activity was then in the form of unchanged LSD in the organs, the remainder consisting of water-soluble transformation products of LSD. Since the peak of the psychic effect occurs only when most of the LSD has disappeared from the organs, it may be concluded with great probability from these experiments that even minimal amounts of LSD are capable of triggering a chain of reactions at the end of which the psychic symptoms occur.

The interest that LSD and other psychotomimetics have found in experimental psychiatry and psychotherapy is due to the following:

a) The experimental psychosis induced with the psychotomimetics on healthy and mentally ill persons enables the study of the individual psychotic symptoms and the biochemical processes associated with them.

b) The psychiatrist has the opportunity to experience the symptoms and sensations of patients in experimental psychosis on himself.

c) Pharmaceuticals can be tested for their antipsychotic effect on the model psychosis.

d) The discovery of such a highly potent substance as LSD, which is capable of producing schizophrenia-like states even in trace amounts, stimulated research into finding biochemical factors in psychoses that have no underlying organic cause. The discovery of a substance as the cause of mental illness would open up new possibilities for its chemotherapeutic treatment.

e) The psychic loosening and the activation of repressed contents of consciousness, which are produced by the psychotomimetics, facilitate psychoanalysis and improve the chances of success of a psychotherapeutic treatment.

A. K. Busch and *W. C. Johnson* in the USA were the first to use LSD as a medicinal aid in psychotherapy[332]. In the following years publications about this application appeared by *H. A. Abramson*[349] in the USA, *R. A. Sandison*[333] in England, *W. Frederking*[334] in Germany, and many other authors, in which the shortening of the treatment time and the successful treatment of cases resistant in the ordinary psychotherapy are emphasized. A more recent survey of this field can be found in *H. Leuner* and *H. Holfeld*[335], who have made special contributions to the development of the technique of this medicinal psychotherapy, which they call psycholytic treatment.

Modification products of LSD

In order to determine the relationship between chemical structure and psychic activity in the group of lysergic acid amides, the molecule of LSD was modified and substituted in various ways. These derivatives, whose preparation and properties are described in Section C, VII, were subjected on the one hand to a more or less comprehensive pharmacological analysis in animal experiments and on the other hand a selection of them was tested for psychic activity in humans. For the comparison between pharmacological properties and psychic efficacy, the findings of 18 typical alteration products were selected and schematically presented by *A. Cerletti*[312] (Figure 25).

On the left side, the psychotomimetic activity is plotted logarithmically in relative values, with regard to LSD = 100. These values originate for the most part from investigations of *H. Isbell* in the USA[336]; in part they were determined in self-experiments[337]. It can be seen from this representation that when the diethylamide group is replaced by other amide groupings, the psychic effect of LSD is still present, but always in a weakened form. Even the closely related dimethylamide has only about 1/10 of the potency of LSD.

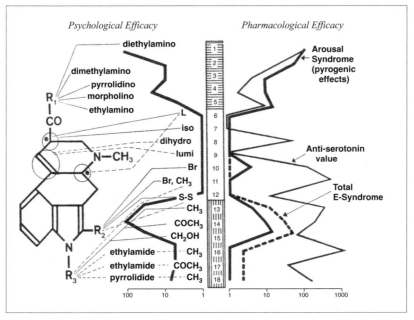

Figure 25. Comparison of the effects of LSD and related substances
Psychological efficacy and pharmacological efficacy are given in logarithmically
plotted relative values (LSD = 100).
Anti-serotonin value = anti-serotonin effect.
The numbers 1–18 in the center of the figure correspond to the arbitrary num-
bering of the compounds studied (see table below).

Table for Figure 25

No.	Lysergic acid derivative	No.	Lysergic acid derivative
1	Lysergic acid diethylamide (LSD)	10	2-Bromo-LSD (BOL-148)
2	d-Lysergic acid dimethylamide	11	1-Methyl-2-bromo-LSD
3	d-Lysergic acid pyrrolidide	12	Di-LSD-disulfide
4	d-Lysergic acid morpholide	13	1-Methyl-LSD
5	d-Lysergic acid monoethylamide	14	1-Acetyl-LSD
6	l-Lysergic acid diethylamide	15	1-Hydroxymethyl-LSD 1-Methyl-d-lysergic acid
7	d-Isolysergic acid diethylamide	16	monoethylamide
8	Dihydrolysergic acid-(I) diethylamide	17	1-Acetyl-d-lysergic acid monoethylamide
9	Lumi-lysergic acid-(I) diethylamide	18	1-Methyl-d-lysergic acid pyrrolidide

Compounds 1-5 represent variations on the acid amide moiety of lysergic acid. Compounds
6-12 include stereoisomers and various substitutions on the ring skeleton of LSD that have
resulted in psychically ineffective substances. In compounds 13-18, the indole nitrogen of
lysergic acid is substituted with various residues.

The psychic efficacy has been practically completely lost in the stereoisomers of LSD, furthermore in the derivatives in which the double bond in ring D has been saturated with hydrogen or by addition of water. Also practically ineffective are the derivatives with a substituent in the 2-position, e.g., the 2-bromo-LSD (known under the experimental number BOL-148), or the bimolecular product linked with an S-S bridge in the 2-position.

Among the derivatives with substituents at the indole nitrogen, however, compounds with considerable psychic activity are found again. Thus, 1-acetyl-LSD is still as potent as LSD. Combinations of substitutions at the amide group and in the 1-position always resulted in attenuated compounds with activities below 10% of LSD. Among all the many relatives of LSD, no substance has yet been found that surpasses it in potency.

On the right side of the diagram, the pharmacological efficacy is plotted logarithmically. As an expression of the pharmacological efficacy the excitation syndrome was chosen, as it appears in LSD mainly as a result of stimulation of sympathetic centers, i.e., mydriasis, piloerection, temperature increase, etc. In particular, the temperature increase in rabbits gives a good measure of the central-vegetative stimulation.

Where the pyrogenetic effect parallels the general excitation syndrome, namely in compounds 1–9, the solid line applies. However, substitution in the 1-position attenuates the pyrogenetic effect in comparison with the other sympathetic stimulus symptoms, so that an average of the total arousal syndrome is plotted here as a dashed line. Also in compound 10, bromo-LSD, there is a discrepancy between temperature increase and excitation syndrome. Bromo-LSD causes only a slight increase in temperature, but does not cause any symptoms of central vegetative stimulation.

On another curve are plotted the values of antiserotonin activity, also a characteristic effect of LSD.

After the discovery of the high activity of LSD as a serotonin blocker[356, 357], it was hypothesized that the psychic effects of LSD could be result from its blocking serotonin in the brain. However, comparison of the curve of anti-serotonin efficacy with that of psychic efficacy clearly shows that this hypothesis cannot be correct. Compounds such as bromo-LSD or, to an even greater extent, 1-methyl-2-bromo-LSD, but also 1-methyl-lysergic acid ethylamide, in some cases surpass LSD several times over in terms of anti-serotonin activity, but are psychically only weak or practically completely ineffective.

A different picture emerges if one compares the curve of the arousal syndrome with that of the psychic efficacy. Here one finds an extensive parallelism of the activities. The most psychically effective compounds, LSD and acetyl-LSD, show maximum values on both curves, and the series of compounds which are psychotomimetically inactive (compounds 6–12) prove to be hardly effective even when measured by the arousal syndrome.

From this correlation, the following conclusion can be drawn as a result of these comparative psychopharmacological studies: in LSD and its relatives, there is a connection between psychotomimetic efficacy on the one hand and the pharmacological effect of extensive central sympathicotonic stimulation on the other, which can be explained by stimulation of sympathetic centers in the hypothalamus.

2-Bromo derivatives

In addition to variation of the amide side chain and hydrogenation of the double bond at the 9–10 position, substitution of the 2-position by halogen and alkylation at the indole nitrogen have led to pharmacologically interesting modifications of ergot alkaloids, the most important representatives of which are discussed below.

Effect spectra of:

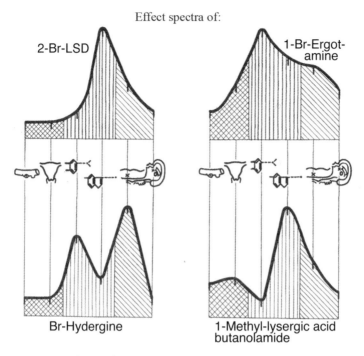

Figure 26. For the explanation of the signs between the curves, see Figure 21. Br-Hydergine = mixture of equal parts of 2-bromo-dihydroergocristine, 2-bromo-dihydroergocryptine, and 2-bromo-dihydroergocornine in the form of the methanesulfonates.

Figure 26 shows the activity spectra of bromo-LSD, bromo-ergotamine and the mixture of bromo derivatives of dihydroergocristine, -ergocryptine, and -ergocornine (= Hydergine), from which the strong shift in the relative strength of the active components compared to the non-brominated compounds is evident.

The psychotomimetic efficacy of LSD has practically completely disappeared in the case of the bromo derivative, as can already be seen from Figure 25, where bromo-LSD is listed as substance 10. In the pharmacological action pattern, this is reflected in the recession of the excitation syndrome caused by stimulation of central sympathetic structures. Bromo-LSD (BOL-148), however, is still as effective as LSD in terms of its serotonin-blocking effect[257-260, 338]. The adrenolytic effect component and the smooth muscle effects recede completely. Up to now, bromo-LSD has found only experimental application.

In the case of ergotamine, bromination affected the modification of the action spectrum as shown in Figure 26. The adrenaline inhibitory effect of bromo-ergotamine is enhanced up to twofold compared to that of ergotamine, depending on the test, whereas the uterotonic effect is very much attenuated. In certain tests, e.g., on rabbit uterus in situ, it is no longer detectable at all. The vasoconstrictor, blood pressure-increasing component of the ergotamine action is very much attenuated in bromo-ergotamine. The central effects also recede, as evidenced by a 5–10 times lower toxicity compared to ergotamine[339].

The introduction of bromine into the dihydro derivatives of the peptide alkaloids results in shifts in action, which can be seen from the spectrum of bromo-hydergine (Figure 26). The pronounced adrenolytic action of hydergine is greatly attenuated. In contrast, the central antihypertensive effect of the parent substance remains unaffected. Other central effects, such as the emetic activity, are less pronounced. The influence on central sympathetic structures is strongly attenuated, which is reflected in reduced toxicity[339].

1-Alkyl derivatives; 1-methyl-*d*-lysergic acid (+)-butanolamide-(2)

When the hydrogen atom on the indole nitrogen is replaced by an alkyl residue, especially by methyl, the serotonin antagonistic effect component in almost all lysergic acid derivatives is selectively enhanced, usually several times over. Since certain lysergic acid derivatives, especially LSD, are in and of themselves among the most effective serotonin antagonists, the methylation of these compounds has led to extraordinarily active, specific serotonin blockers. Serotonin antagonists play an important role in contemporary pharmacological research because they can be used to study the biological functions of serotonin[325]. Although these have by no means been elucidated, the results available to date show that serotonin is an endogenous agent that significantly influences important structures and functions of the organism, so that substances with a specific serotonin antagonistic effect are of great interest from both a theoretical-experimental and a practical-therapeutic point of view.

Table 49

Enhancement of antiserotonin effect on isolated rat uterus by 1-methylation of lysergic acid.

	R	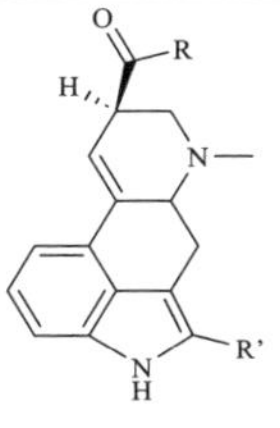	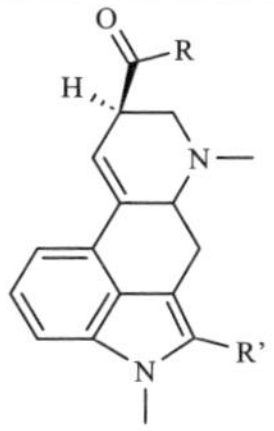
Lysergic acid diethylamide (LSD)	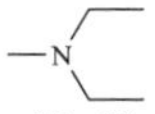 —N⟨ (R' = H)	100.0	370
2-Bromo-lysergic acid diethylamide	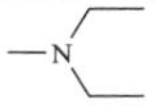 —N⟨ (R' = Br)	100.0	530
Lysergic acid ethylamide	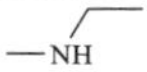 —NH (R' = H)	12.0	835
Lysergic acid pyrrolidide	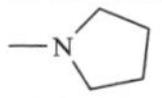 —N◯ (R' = H)	4.7	130
Lysergic acid amide	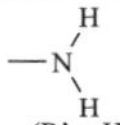 —N⟨H,H (R' = H)	4.3	160
Ergobasine	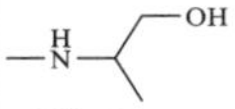 —N(H)—⟨—OH (R' = H)	17.0	400
Methergine	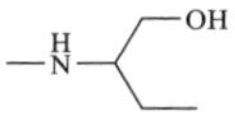 —N(H)—⟨—OH (R' = H)	60.0	250

Test objects for testing the serotonin inhibitory effect are mainly the isolated rat uterus[340] and, as an in vivo test, the serotonin edema in the rat paw[341].

Table 49 shows the increase in antiserotonin effect on rat uterus upon methylation of various lysergic acid derivatives[342].

The increase is particularly strong by compounds which are in some cases 20 times less potent than LSD and which, after methylation, exceed it several times over, such as lysergic acid ethylamide. The introduction of the methyl residue in the natural alkaloid ergobasine has a very striking effect, in that the relative activity based on LSD as a standard (100%) increases from 17% to 400%. The semisynthetic homologue of ergobasine, lysergic acid butanolamide, oxytocicum known as Methergine® (aka: Methylergometrine or methylergonovine), also undergoes a multiple increase in potency.

A somewhat different picture emerges when the influence of methylation on serotonin antagonism is compared to the inhibition of serotonin edema in the rat paw[341]. One then receives results that are summarized in Table 50.

1-Methyl-LSD and 1-methyl-2-bromo-LSD, which are several times more potent than their parent substances in the rat uterus test, are somewhat less effective on the edema test than LSD and bromo-LSD, respectively. In contrast, d-lysergic acid monoethylamide experiences an increase in efficacy to about twice that. The same applies to ergotamine, which itself, however, is only weakly effective. The influence of methylation on the two known oxytocics, d-lysergic acid L-propanolamide-(2) (ergobasine) and d-lysergic acid (+)-butanolamide-(2) (Methergine®), is remarkable. For ergobasine, the potency increases from half that of LSD to about 2½-fold LSD potency, and for methergine from 1½-fold to 4½-fold potency. This different influence of methylation on the antiserotonin activity, measured by the edema test, is evident from Figure 27[342].

1-Methyl-d-lysergic acid butanolamide-(2), which has been given the generic name designation Methysergide and introduced into therapy under the brand name Deseril®, is the most potent serotonin inhibitory agent, according to in vivo and in vitro testing. It antagonizes not only peripheral but also central effects of serotonin. Thus, it inhibits its barbiturate-potentiating effect, but not that of chlorpromazine, which is again a sign of the specificity of its serotonin antagonism. The specificity of serotonin inhibition is also documented by the fact that it takes a 9,200 times higher concentration of Methysergide to inhibit acetylcholine action to the same degree in the same organ (rat uterus).

Table 50

Antiserotonin efficacy in rat paw edema assay,
of lysergic acid derivatives

Group	Substance	ED50 (Base) in μg/kg	% relative efficacy (LSD = 100)
LSD derivatives	d-Lysergic acid diethylamide (LSD)	56.8	100.0
	9, 10-Dihydro-LSD	470.0	12.1
	1-Methyl-LSD	62.2	91.3
	2-Bromo-LSD	196.5	28.9
	1-Methyl-2-bromo-LSD	218.0	26.1
Other acid amides of lysergic acid	d-Lysergic acid dimethylamide	455.0	12.5
	d-Lysergic acid monoethylamide	257.0	22.1
	1-Methyl-d-lysergic acid mono-ethylamide	96.8	58.7
Ergotamine group	Ergotamine	844.0	6.7
	Dihydroergotamine	833.0	6.8
	1-Methyl-ergotamine	430.0	13.2
Ergobasine group	d-Lysergic acid L-propanolamide-(2) (= Ergobasine)	122.0	46.8
	9,10-Dihydroergobasine	895.0	6.4
	1-Methyl-ergobasine	21.9	259.4
Methergine group	d-Lysergic acid (+)-butanolamide-(2) (= Methergine)	37.4	151.9
	9,10-Dihydromethergine	271.0	21.0
	1-Methyl-methergine	12.9	440.3

It is further evident in the circulatory system and in the reduction of over-all serotonin toxicity[343]. The antiserotonin effect of methysergide is highly specific, as can be seen from the shape of the curve of its action spectrum (see Figure 26). The oxytocic effect is very low, both in rabbits and in humans. Adrenolytic effects are completely absent. Also absent are the vasoconstrictor and pressor effects so typical of natural ergot alkaloids. In humans, Methysergide has no psychotomimetic effects. This corresponds pharmacologically to the absence of the centrally triggered excitation syndrome that characterizes the action spectrum of LSD. Related to this is also the low toxicity of Methysergide, which is about 10 times less than that of the nonmethylated *d*-lysergic acid (+)-butanol-amide-(2).

Therapeutically, Methysergide (Deseril®) is used in clinical disorders in which serotonin seems to play a physiopathological role. In certain cases of carcinoid, it can bring about a marked reduction in the troublesome symptoms (diarrhea, flushing, dyspnea)[344]. Furthermore, it exhibits an impressive efficacy in the interval treatment of otherwise therapy-resistant headaches of the vascular type[345–348]. Therapeutic possibilities for this specific serotonin antagonist[343] also exist in the field of rheumatic diseases, allergy, and peripheral circulatory disorders.

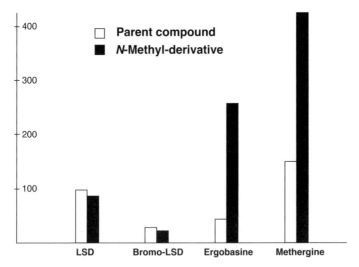

Figure 27. Serotonin inhibition edema test in the rat paw (relative efficacy based on LSD = 100).

The molecule of ergot alkaloids has been modified in many directions over the last 20 years since the partial synthetic preparation of ergobasine and other simple acid amide-like derivatives of lysergic acid as part of studies to modify their pharmacological properties. The example of the serotonin antagonist Methysergide, which has only recently been developed, shows that even after so many years of research in this field, it was still possible today to work out new qualities of action from the uniquely pluripotent basic structure of the ergoline ring scaffold. With the newly developed possibility of producing the peptide part with its novel structural elements totally synthetically and thus also chemically modifying and varying the ergot alkaloids of the ergotamine-ergotoxine group, a wide and interesting field of research has opened up again for the pharmaceutical chemist. Thus, the chemically fascinating substance class of ergot alkaloids is still not exhausted as a source of pharmacologically active substances that can lead to valuable medicines.

AFTERWORD

Many of us recall Albert Hofmann with the greatest fondness and respect, as the extraordinary man he was: chemist, visionary, philosopher, and devoted husband and father. His passing was widely mourned throughout the world.

Ergot Alkaloids (*Die Mutterkornalkaloide*) was Albert's seminal tribute to the classical group of nitrogen-containing metabolites of the fungus *Claviceps purpurea*, Clavicipitaceae including ergotamine and the ergotoxines, all derived from the same intermediate (lysergic acid) and then linked to a cyclic tripeptide. The story of how this landmark work came to be is a fascinating one.

Albert's first synthesis of LSD was in 1938. That day he neglected to have lunch with colleagues, and ate alone in his lab, after which he felt a lightness of being and minor effects later attributable to a low, subthreshold dose. The Bicycle Day event on April 19, 1943, occurred after a purposeful self-administration of 250 micrograms, unwittingly several times a threshold dose, followed by a two-and-a-half-mile ride home, cycling without a car during wartime.

The Bicycle Day dose was so extraordinary that Albert thought he may have inaccurately weighed the LSD. Three colleagues agreed, including Ernst Rothlin, director of the pharmacology department at Sandoz. They self-administered LSD, using less than one-third of Albert's dose of 250 micrograms. That 80 micrograms, well past the threshold dose, was sufficient to convince them otherwise.

After Albert's first synthesis of LSD in 1938 and the famed Bicycle Day event in 1943, followed by its scheduling in California in 1966 and the United States in 1968, Albert decided to write about the ergot alkaloids with a focus on their chemical structure and pharmacology. Well-received by the scientific community, it quickly became the standard reference work on the topic. *Ergot Alkaloids* is significant for its detailed analysis of structure and activity, demonstrating natural products' potential for new medicines, many compounds of which are evolving through the application of artificial intelligence to drug discovery today.

Albert would be delighted at the expansion of worldwide production of his beloved ergot alkaloids, several of which he created. The industrial producers are the major firms Sanofi in France (Cafergot and Dihydergot for migraine), Taj Pharmaceuticals in India (Ergamisol and Ergodryl), and Gedeon Richter in Hungary (Ergoset for Parkinson's disease and Albert's Hydergine to improve cognitive function).

The world production of ergot alkaloids now reaches thousands of kilograms annually, including both ergopeptines and semisynthetic ergot alkaloid derivatives (e.g., cabergoline and pergolide). In 2010 alone, the total world production was more than 20,000 kilograms, with the field cultivation of C. *purpurea* contributing about 50 percent and the balance often through submerged culture leading to paspalic acid.

Albert conceived of the structure of LSD while pondering the circulatory stimulant Coramine, which is nicotinic acid diethylamide, and planned to make the analogous diethylamide of lysergic acid in hope of obtaining a novel circulatory medicine.

The worldwide figures for LSD production, which since 1966 remains almost entirely clandestine, amount to several 100 MM doses annually. Almost all is derived from lysergic acid obtained from hydrolysis of the classic ergopeptines: ergotamine, ergocristine, ergocryptine, and ergocornine.

Yet this scholarly volume, *Ergot Alkaloids*, has an exceptional human story, and it is that of Albert Hofmann.

Hofmann's first memory was of large red strawberries in his family's garden. At four years of age, with his mother, he saw a comet in the sky: Halley's Comet of 1910. At eight, he could hear the guns in the Alsace in World War I.

In his boyhood, Albert often played near the ruins of a Habsburg castle, the Stein. On a May morning, he was wandering on a forest path near Martinsburg above Baden, reveling in his childhood freedom and natural mind. Suddenly imbued with awareness of the oneness of nature, he experienced an epiphany not unlike the transcendence he found during his later experiences with LSD.

"As I strolled through the freshly greened woods filled with birdsong and lit up by the morning sun, all at once everything appeared in an uncommonly clear light. It shone with the most beautiful radiance, speaking to the heart, as though it wanted to encompass me in its majesty. I was filled with an indescribable sensation of joy, oneness, and blissful security."

In 1929, at the age of only 23, he would be awarded a PhD from the University of Zurich for his dissertation on vineyard snails and how they digest the carbohydrate chitin, of which insect exoskeletons are composed.

Albert met Anita Guanella in 1934 while on a skiing holiday, and they married later that year. They had two sons, then two daughters: Dieter (1936–1989), Andreas (1939), Gaby (1942–2007), and Beatrix (1948). They ascended the Bernina Peak together and traveled to India, Thailand, and Huautla de Jiménez in Oaxaca, Mexico. Devoted to each other, Anita and Albert were married for 74 years.

In 1955, Albert became friends with Aldous Huxley and provided him with LSD in response to Aldous's request for a sample for his research. Albert retained a letter from Laura Archera Huxley in 1963 asking for 150 micrograms of LSD to alleviate the pain from Aldous's laryngeal cancer. Prior to Aldous's death on November 22, 1963, Laura administered Albert's LSD to him in response to his final words in handwriting: "LSD, 100 micrograms, I.M." (intramuscular).

During the early arc of his career, Albert bicycled to Sandoz, returning home and often working in his study to answer correspondence by mail from throughout the world, but never mentioned to his children his discoveries or his prominence. His son Andreas only learned of Albert's global recognition when he visited New York City as a young man.

Other than his explorations in chemistry, Albert was an avid mountaineer, hiking and climbing in the Swiss Alps. He played the violin and piano and was an accomplished woodcarver as well as a passionate botanist. He spoke English, French, German, Italian, Romansh, and Portuguese. An avid reader, he studied psychology, philosophy, and Eastern spirituality.

Psychedelic and academic literature is replete with references to Albert's accomplishments, not only the first synthesis of LSD, but also the first inadvertent and planned exposures to LSD, and the first observation of its subjective effects. Albert received samples of the fabled Mexican mushroom (*Psilocybe mexicana*, Strophariaceae) and identified lysergic amides underlying the hallucinogenic properties of *Ololiuqui*, the psychoactive morning glories (*Ipomoea* spp, *Rivea corymbose* etc., Convolvulaceae).

He conducted the first syntheses and naming of psilocybin, psilocin, and many of their analogues. Having prepared pure synthetic psilocybin, he provided it to the Mexican curandera Maria Sabina to use in a ceremony. Although Maria utilized *P. mexicana* in her mushroom healing veladas among the Mazatec, Maria pronounced that Albert's psilocybin was "the same."

In later years, Albert became an advocate for sustainable living and environmental protection. With their love of nature, Albert and Anita moved their home three times, always further into the forest. Albert was so enamored by a charming place near Burg im Leimental in the Rittematte area of the Swiss Jura mountains that he bought the property on sight. It was a plot of land which straddled the border between Switzerland and France. With only a stroll at Albert and Anita's place, one could be in either country.

On a verdant slope overlooking the town, Albert and his grandson carved a bench, allowing grand, serene views of the mountains. In the afternoons, Albert and Anita would sit there and watch the sunsets together. They had two trees, one for each other, near the bench they so loved.

Anita and Albert were fond of serving guests a glass of *Pflaumenschnaps* or *Kirschwasser* (plum or cherry schnapps) from fruit Albert had fermented and distilled in his cellar in a large glass apparatus from his Sandoz laboratory. Albert even sketched the labels on the bottles and grew the cherry trees.

In addition to Aldous Huxley, Albert's many friends included Humphrey Osmond, Allen Ginsberg, Timothy Leary, Gordon Wasson, and Alexander (Sasha) Shulgin. Sasha was fond of explaining to Albert that his favorite drug was LSD, whereupon Albert would reply that his favorite drug was MDMA.

The surreal artist H.R. Giger was also a friend. Albert and Giger designed a stained-glass window for St. Martin's Church in Grueningen, Switzerland, depicting a cosmic Christ surrounded by psychedelic imagery.

At the 1996 Worlds of Consciousness Conference in Heidelberg, Albert and I met for the first time. Albert, then in his 90s, appeared at 3 a.m. in his gray suit, only to be surrounded by masses of colorful students (some with green hair) who listened attentively to his every word. Three years later, he was nominated for the title of the "most influential person of the 20th century" by readers of the German newspaper *Die Welt*. In 2007, he tied for first place in a list of the 100 greatest living geniuses published by the *Daily Telegraph*.

Albert celebrated his 100th birthday by ingesting a small quantity of LSD, reporting a renewed connection to the natural world. The next year, he was nominated for the Nobel Prize in Chemistry for the third time.

Anticipating death, he remarked, "I go back to where I came from, to where I was before I was born, that's all."

Albert wrote his own eulogy, describing his love of the meadows and flowers he saw as a boy.

"Paracelsus described nature and creation as a 'book that was written by God's finger.' During my life I was given this exhilarating and entirely comforting experience: The one who understands how to read this book, not only with regards to scientific research but with marveling and loving eyes, will find a deeper, wonderful reality revealing itself—a reality in which we are secure and united for ever and ever."

Predeceased by Anita, Albert died on April 29, 2008, at the age of 102, leaving behind a global legacy of understanding, compassion, and scientific inquiry. Albert and Anita are each buried under their tree. Albert's last words were a profound understatement:

"It's all been very interesting."

William Leonard Pickard
Research Affiliate, Harvard Law School
Petrie-Flom Center for Health Law Policy, Biotechnology & Bioethics,
Program on Psychedelics Law and Regulation
May 2023

BIBLIOGRAPHY

[1] *Stoll, A.,* Fortschr. Chem. Org. Naturstoffe [Wien] 9, 114–174 (1952).

[2] Ibid. Progr. Allergy 3: 388 (1952).

[3] *Glenn, A. L.,* Quart. Rev. (Chem. Soc., London) 8, 192–218 (1954).

[4] *Guggisberg, H.,* Mutterkorn. Vom Gift zum Heilstoff. S. Karger, Basel–New York, 1954.

[5] *Grasso, V.,* Le Claviceps delle Graminacee Italiane (The Claviceps of the Italian Gramineae). Ann. Sperimentaz. Agrar. [Roma] N. S. 6, 747 (1952).

[6] *Barger, G.,* Ergot and Ergotism. Gurney and Jackson, London, 1931.

[7] *Fuchs, L.,* and *Pöhm, M.,* Scientia Pharmac. 21, 239 (1953).

[8] *Langdon, R. F. N.,* University of Queensland Papers, Dept. Botany 3, 61 (1954).

[9] *Walker, J.,* Proc. Linn. Soc. N. S. W. 82, 322 (1957).

[10] *Pantidou, M. E.,* Can. J. Botany 37, 1233 (1959).

[11] *Kawarani, T.,* Bull. Nat. Hyg. Lab. Tokyo 70, 127 (1952); 71, 161 (1953).

[12] *Grasso, V.,* Ann. Sperimentaz. Agrar. [Roma] N.S. 8, 1 (1954).

[13] Ibid. Boll. Staz. Patol. Vegetale 15 (Series 3) 317 (1957).

[14] *Lindquist, J. C.,* and *Carranza, J. M.,* Rev. Fac. Agr. [Argentina] 36, 151 (1960).

[15] *Brady, L. R.,* Lloydia 25, 1 (1962).

[16] *Silber, A.,* and *Bischoff, W.,* Arch. Pharmaz. 288, 124 (1955).

[17] *Stoll, A., Brack, A., Kobel, H., Hofmann, A.,* and *Brunner, R.,* Helv. Chim. Acta 37, 1815 (1954).

[18] *Hofmann, A., Brunner, R., Kobel, H., Brack, A.,* Helv. Chim. Acta 40, 1358 (1957).

[19] *Gröger, D., Tyler Jr., V. E.,* and *Dusenberry, J.E.,* Lloydia 24, 97 (1961).

[20] *Arcamone, F., Bonino, C., Chain, E. B., Ferretti, A., Pennella, P., Tonolo, A.,* and *Vero, L.,* Nature [London] 187, 238 (1960).

[21] *Kybal, J.,* and *Brejcha, V.,* Pharmazie 10, 752 (1955).

[22] *v. Békésy, N.,* Phytopathol. Z. 26, 49 (1956).

[23] *Campbell, W. P.,* Can. J. Bot. 35, 315 (1957).

[24] *Meinicke, R.,* Flora [Jena] 143, 395 (1956).

[25] *Abe, M.,* Ann. Rep. Takeda Res. Lab. 10, 73 (1951).

[26] Ibid. Ann. Rep. Takeda Res. Lab. 10, 129 (1951).

[27] *Abe, M., Yamano, T., Kozu, Y.,* and *Kusumoto, M.,* J. Agric. Chem. Soc. [Japan] 25, 458 (1952).

[28] Schweiz. Patent Nr. 321 323 on 10.4.1953, with additional patent Nr. 330 722, Sandoz A. G., Basel.

[29] *Rochelmeyer, H.,* Dtsch. Apotheker-Ztg. 94, 1 (1954).

[30] *Kaiser, H.,* and *Thielmann, E.,* Arch. Pharmaz. 289, 378 (1956).

[31] *Rassbach, H., Büchel, K. G.,* and *Rochelmeyer, H.,* Arzneimittel-Forschg. *6,* 690 (1956).

[32] *Taber, W. A.,* and *Vining, L. C.,* Can. J. Microbiol. *3,* 55 (1957).

[33] Ibid., Can. J. Microbiol. *4,* 611 (1958).

[34] *Rochelmeyer, H.,* Pharmaz. Ztg. *103,* 1269 (1958).

[35] *Kobel, H., Brunner, R.,* and *Brack, A.,* Experientia [Basel] *18,* 140 (1962).

[36] *Arcamone, F., Chain, E. B., Ferretti, A., Minghetti, A., Pennella, P., Tonolo, A.,* and *Vero, L.,* Proc. Roy. Soc. [London], Ser. B., *155,* 26 (1961).

[37] *v. Békésy, N.,* Zbl. Bakteriol. II, *99,* 321 (1938).

[38] *Stoll, A.,* and *Brack, A.,* Pharmac. Acta Helvetiae *19,* 118 (1944).

[39] Ibid., Ber. Schweiz. Bot. Ges. *54,* 252 (1944).

[40] *v. Békésy, N.,* Pharmaz. *11,* 339 (1956).

[41] *Gröger, D.,* Die Kulturpflanze (The Cultivated Plant), Akademie-Verlag, Berlin, S. 226, 1956.

[42] *Mothes, K.,* and *Silber, A.,* Forsch. and Fortschr. [Berlin] *28,* 101 (1954).

[43] *Silber, A.,* and *Bischoff, W.,* Pharmazie *9,* 46 (1954).

[44] *Kobert, R.,* Historische Studien aus dem pharmakologischen Institut der kaiserlichen Universität Dorpat (Historical Studies From the Pharmacological Institute of the Imperial University of Dorpat), *1,* 1–47 (1889), Halle a. S.

[45] *Durrer, R.,* Mitt. Antiquarischen Ges. Zürich *24,* 233 (1898).

[46] *Fülöp-Miller, R.,* Kampf gegen Schmerz and Tod. (Fighting Pain and Death) S. 5, Südost-Verlag, Berlin, 1938.

[47] *Mellanby, E.,* Brit. Med. J. *1,*(3614), 677. (1930).

[48] *Stearns, J.,* Med. Repository N.Y. *5,* 308 (1808).

[49] *Hosack, D.,* Observations on ergot. Essays on various subjects of medical science. Vol. 2, pp. 295-301, J. Seymour, New York (1824).

[50] *Vauquelin, M.,* Ann. Chim. (Phys.) *3,* 337 (1816).

[51] *Tanret, Ch.,* C.R. Acad. Sci., Paris *81/2,* 896 (1875).

[52] *Barger, G.,* and *Carr, F. H.,* J. Chem. Soc. [London] *91,* 337 (1907).

[53] *Kraft, F.,* Arch. Pharm. *244,* 336 (1906).

[54] Ibid. Arch. Pharm. *245,* 644 (1907).

[55] *Dale, H. H.,* J. Physiology *34,* 163 (1906).

[56] *Stoll, A.,* Verh. Schweiz. Naturf. Ges. *1920,* 190.

[57] Schweiz. Patent Nr. 79 879 (1918); D.R.P. 357 272 (1922), Sandoz A.G., Basel.

[58] *Stoll, A.,* Schweiz. Apotheker-Ztg. *60,* 341 (1922).

[59] Ibid. Schweiz. Apotheker-Ztg. *60,* 358 (1922).

[60] Ibid. Schweiz. Apotheker-Ztg. *60,* 374 (1922)

[61] Ibid. Helv. Chim. Acta *28,* 1283 (1945).

[62] *Spiro, K.,* and *Stoll, A.,* Schweiz. Med. Wschr. *51,* 525 (1921).

[63] *Barger, G.,* and *Carr, F. H.,* Chem. News *94,* 89 (1906).

[64] *Dudley, H. W.,* and *Moir, C.,* Brit. Med. J. *1,* 520 (1935).

[65] *Stoll, A.,* and *Burckhardt, E.,* C. R. Acad. Sci. Paris *200,* 1680 (1935).

[66] Ibid., Bull. Sci. Pharm. *42,* 257 (1935).

[67] *Kharasch, M. S.*, and *Legault, R.R.*, Science [Lancaster] *81*, 388, 614 (1935).

[68] *Thompson, M. R.*, Science [Lancaster] *81*,636 (1935).

[69] *Smith, S.*, and *Timmis, G. M.*, J. Chem. Soc. [London] *1937*, 396.

[70] *Stoll, A.*, and *Burckhardt, E.*, Hoppe-Seyler's Z. Physiol. Chem. *250*, 1 (1937).

[71] *Stoll, A.*, and *Hofmann, A.*, Helv. Chim. Acta *26*, 1570 (1943).

[72] *Abe, M.*, and *Yamatodani, S.*, J. Agric. Chem. Soc. [Japan] *28*, 501 (1954).

[73] Ibid., Bull. Agric. Chem. Soc. [Japan] *19*, 161 (1955).

[74] *Abe, M.*, *Yamatodani, S.*, *Yamano, T.*, and *Kusumoto, M.*, Bull. Agric. Chem. Soc. [Japan] *20*, 59 (1956).

[75] *Abe, M.*, *Yamano, T.*, *Yamatodani, S.*, *Kozu, Y.*, *Kusumoto, M.*, *Komatsu, H.*, and *Yamada, S.*, Bull. Agric. Chem. Soc. [Japan] *23*, 246 (1959).

[76] *Yamatodani, S.*, Ann. Rep. Takeda Res. Lab. *19*, 1 (1960).

[77] *Abe, M.*, *Yamatodani, S.*, *Yamano, T.*, and *Kusumoto, M.*, Agric. Biol. Chem. [Tokyo] *25*, 594 (1961).

[78] *Hofmann, A.*, Planta Med. [Stuttgart] *9*, 354 (1961).

[79] *Spilsbury, J. F.*, and *Wilkinson, S.*, J. Chem. Soc. [London] *1961*, 2085.

[80] *Schlientz, W.*, *Brunner, R.*, *Frey, A. J.*, *Stadler, P. A.*, and *Hofmann, A.*, Helv. Chim. Acta (in Press).

[81] *Smith, S.*, and *Timmis, G. M.*, J. Chem. Soc. [London] *1932*, 763.

[82] *Jacobs, W. A.*, and *Craig, L. C.*, J. Biol. Chemistry *104*, 547 (1934).

[83] Ibid., J. Biol. Chemistry *106*, 393 (1934).

[84] Ibid., Science [Lancaster] *82*, 16 (1935).

[85] Ibid., J. Biol. Chemistry *108*, 595 (1935).

[86] Ibid., J. Biol. Chemistry *110*, 521 (1935).

[87] Ibid., J. Amer. Chem. Soc. *57*, 960 (1935).

[88] *Smith, S.*, and *Timmis, G. M.*, J. Chem. Soc. [London] *1936*, 1440.

[89] *Stoll, A.*, and *Hofmann, A.*, Hoppe-Seyler's Z. Physiol. Chem. *251*, 155 (1938).

[90] *Jacobs, W. A.*, and *Craig, L. C.*, J. Amer. Chem. Soc. *60*, 1701 (1938).

[91] *Uhle, F. C.*, and *Jacobs, W. A.*, J. Org. Chem. *10*, 76 (1945).

[92] *Stoll, A.*, *Hofmann, A.*, and *Troxler, F.*, Helv. Chim. Acta *32*, 506 (1949).

[93] *Stoll, A.*, *Rutschmann, J.*, and *Schlientz, W.*, Helv. Chim. Acta *33*, 375 (1950).

[94] *Stoll, A.*, *Hofmann, A.*, and *Petrzilka, Th.*, Helv. Chim. Acta 34, 1544 (1951).

[95] *Kornfeld, E. C.*, *Fornefeld, E. J.*, *Kline, G. B.*, *Mann, M. J.*, *Jones, R. G.*, and *Woodward, R. B.*, J. Amer. Chem. Soc. *76*, 5256 (1954).

[96] *Kornfeld, E. C.*, *Fornefeld, E. J.*, *Kline, G. B.*, *Mann, M. J.*, *Morrison, D. E.*, *Jones, R. G.*, and *Woodward, R. B.*, J. Amer. Chem. Soc. *78*, 3087 (1956).

[97] *Stoll, A.*, *Petrzilka, Th.*, *Rutschmann, J.*, *Hofmann, A.*, and *Günthard, Hs. H.*, Helv. Chim. Acta *37*, 2039 (1954).

[98] *Leemann, H. G.*, and *Fabbri, S.*, Helv. Chim. Acta *42*, 2696 (1959).

[99] *Stadler, P. A.*, and *Hofmann, A.*, Helv. Chim. Acta *45*, 2005 (1962).

[100] *Hofmann, A., Frey, A. J.,* and *Ott., H.,* Experientia [Basel] *17,* 206 (1961).

[101] *Jacobs, W. A.,* and *Gould jr., R. G.,* J. Biol. Chem. *120,* 141 (1937).

[102] *Pöhm, M.,* Naturwissenschaften *44,* 620 (1954).

[103] *Schlientz, W., Brunner, R., Hofmann, A., Berde, B.,* and *Stürmer, E.,* Pharm. Acta Helvetiae *36,* 472 (1961).

[104] *Schlientz, W., Brunner, R., Thudium, F.,* and *Hofmann, A.,* Experientia [Basel] *17,* 108 (1961).

[105] *Szendey, G. L., Renneberg, K. H.,* and *Hartmann, P.,* Naturwissenschaften *48,* 223 (1961).

[106] *Allport, N. L.,* and *Cocking, T. T.,* Quart. J. Pharm. Pharmacol. *5,* 341 (1932).

[107] *Grant, R. L.,* and *Smith, S.,* Nature [London] *137,* 154 (1936).

[108] *Moir, C.,* Brit. Med. J. I, *1932,* 1119.

[109] *Smith, S.,* and *Timmis, G. M.,* J. Chem. Soc. [London] *1936,* 1166.

[110] Ibid., Nature [London] *136,* 259 (1935).

[111] *Stoll, A.,* and *Hofmann, A.,* Helv. Chim. Acta *33,* 1705 (1950).

[112] *Hofmann, A.,* and *Tscherter, H.,* Experientia [Basel] *16,* 414 (1960).

[113] *Schultes, R. E.,* A Contribution to our Knowledge of *Rivea Corymbosa.* The Narcotic Ololiuqui of the Aztecs. Botanical Museum of Harvard University, Cambridge, Massa- chusetts, 1941.

[114] *Smith, S.,* and *Timmis, G. M.,* J. Chem. Soc. [London] *1934,* 674.

[115] *Abou-Chaar, C. I., Brady, L. R.,* and *Tyler jr., V. E.,* Lloydia 24, 89 (1961).

[116] *Stoll, A.,* and *Hofmann, A.,* Hoppe-Seyler's Z. Physiol. Chem. *250,* 7 (1937).

[117] Ibid., Helv. Chim. Acta *26,* 922 (1943).

[118] Belg. Patent Nr. 609 011; French Patent Nr. 1303 289, Sandoz A.G., Basel.

[119] *Jacobs, W. A.,* and *Craig, L. C.,* J. Biol. Chem. *115,* 227 (1936).

[120] *Stoll, A.,* and *Hofmann, A.,* Helv. Chim. Acta *26,* 2070 (1943).

[121] Ibid., and *Petrzilka, Th.,* Helv. Chim. Acta *29,* 635 (1946).

[122] *Jacobs, W. A.,* J. Biol. Chem. *97,* 739 (1932).

[123] *Jacobs, W. A.,* and *Craig, L. C.,* J. Biol. Chem. *113,* 767 (1936).

[124] Ibid., J. Biol. Chem. *111,* 455 (1935).

[125] Ibid., J. Biol. Chem. 128, 715 (1939).

[126] *Jacobs, W. A., Craig, L. C.,* and *Rothen, A.,* Science [Lancaster] *83,* 166 (1936).

[127] *Craig, L. C., Shedlovsky, Th., Gould Jr., R. G.* and *Jacobs, W. A.,* J. Biol. Chem. *125,* 289 (1938).

[128] *Jacobs, W. A.,* and *Gould jr., R. G.,* J. Biol. Chem. *126,* 67 (1938).

[129] Ibid., J. Biol. Chem. *130,* 399 (1939).

[130] *Gould Jr., R. G., Craig, L. C.,* and *Jacobs, W. A.,* J. Biol. Chem. *145,* 487 (1942).

[131] *Hofmann, A.,* Helv. Chim. Acta *30,* 44 (1947).

[132] *Troxler, F.,* Helv. Chim. Acta *30,* 163 (1947).

[133] *Cookson, R. C.,* Chem. and Ind. *1953,* 337.

[134] Stenlake, J. B., Chem. and Ind. *1953*, 1089.

[135] Alder, K., and Dortmann, H. A., Chem. Ber. *86*, 1544 (1953).

[136] Smith, P. F., and Hartung, W. H., J. Amer. Chem. Soc. *75*, 3859 (1953).

[137] Djerassi, C., and Geller, L. E., J. Amer. Chem. Soc. *81*, 2789 (1959), and previous works.

[138] Cahn,R. S., Ingold, C. K., and Prelog, V., Expcrientia [Basel] *12*, 81 (1956).

[139] Atherton, F. R., Bergel, F., Cohen, A., Heath-Brown, B., and Rees, A. H., Chem. and Ind. *1953*, 1151.

[140] Grob, C. A., and Renk, E., Helv. Chim. Acta *44*, 1531 (1961).

[141] Walker, G. N., and Weaver, B. N., J. Org. Chem. *26*, 4441 (1961).

[142] Stoll, A., and Rutschmann, J., Helv. Chim. Acta *33*, 67 (1950).

[143] Gould Jr., R. G., and Jacobs, W. A., J. Amer. Chem. Soc. *61*, 2891 (1939).

[144] Stoll, A., and Rutschmann, J., Helv. Chim. Acta. *16*, 1512 (1953).

[145] Stoll, A., and Petrzilka, Th., Helv. Chim. Acta *36*, 1125 (1953).

[146] Ibid., Helv. Chim. Acta *36*, 1137 (1953).

[147] Stoll, A., and Rutschmann, J., Helv. Chim. Acta *37*, 814 (1954).

[148] Grob, C. A., and Voltz, J., Helv. Chim. Acta *33*, 1796 (1950).

[149] Uhle, F. C., J. Amer. Chem. Soc. *73*, 2402 (1951).

[150] Barltrop, J. A., and Taylor, D. A.H., J. Chem. Soc. [London] *1954*, 3399, 3403.

[151] Plieninger, H., Chem. Ber. *86*, 25 (1953).

[152] Ibid. Chem. Ber. *86*, 404 (1953).

[153] Ibid. Chem. Ber. *88*, 370 (1955).

[154] Plieninger, H., and Suehiro, T., Chem. Ber. *88*, 550 (1955).

[155] Plieninger, H., and Werst, G., Chem. Ber. 89, 2783 (1956).

[156] Stoll, A., Petrzilka, Th., and Rutschmann, J., Hclv. Chim. Acta *35*, 1249 (1952).

[157] Stoll, A., and Hofmann, A., Helv. Chim. Acta *26*, 944 (1943).

[158] Stoll, A., Petrzilka, Th., and Rutschmann, J., Helv. Chim. Acta *33*, 2254 (1950).

[159] Jacobs, W. A., and Craig, L. C., J. Org. Chem. *1*, 245, (1936).

[160] Stoll, A., Hofmann, A., and Becker, B., Helv. Chim. Acta *26*, 1602 (1943).

[161] Jacobs, W. A., and Craig,L. C., J. Biol. Chem. *122*, 419 (1938).

[162] Barger, G., Handbuch der experimcntellen Pharmakologie, Erg. Werk 6, 84, 222 (1938).

[163] Stoll, A., Petrzilka, Th., and Becker, B., Helv. Chim. Acta *,13*, 57 (1950).

[164] Barger, G., and Ewins, A. J., J. Chem. Soc. [London] *97*, 284 (1910).

[165] Stoll, A., Hofmann, A., Leemann, H. G., Ott, H., and Schenk, H. R., Helv. Chim. Acta *39*, 1165 (1956).

[166] Wrinch, D. M., Nature [London] *137*, 411 (1936).

[167] Ibid. Nature [London] *138*, 241 (1936).

[168] Shemyakin, M. M., Tchaman, E. S., Denisova, L. I., Ravdel, G. A., and Rodionow, W. J., Bull. Soc. Chim. [France] *1959*, 530.

[169] Antonov, W. K., Ravdel, G. A., and Shemyakin, M. M., Chimia [Zürich] *14*, 374 (1960).

[170] French Patent Nr. 1 308 758, Sandoz. A.G., Basel.

[171] *Hofmann, A., Ott, H., Griot, R., Stadler, P.A.,* and *Frey, A. J.,* Helv. Chim. Acta *46,* 2306 (1963).

[172] *Abe, M.,* Ann. Rep. Takeda Res. Lab. *10,* 151 (1951).

[173] *Yui, T.,* and *Takeo, T.,* Jap. J. Pharmacol. *7,* 157 (1958).

[174] *Yamatodani, S.,* and *Abe, M.,* Bull. Agric. Chem. Soc. [Japan] *20,* 95 (1956).

[175] *Schreier, E.,* Helv. Chim. Acta *41,* 1984 (1958).

[176] *Stoll, A., Hofmann, A.,* and *Schlientz, W.,* Helv. Chim. Acta 32, 1947 (1949).

[177] *Abe, M., Yamatodani, S., Yamano, T.,* and *Kusumoto, M.,* Bull. Agric. Chem. Soc. [Japan] *19,* 92 (1955).

[178] *Semonský, M., Beran, M.,* and *Macek, K.,* Collect. Czechoslov. Chem. Commun. *23,* 1364 (1958).

[179] *Abe, M.,* Ann. Rep. Takeda Res. Lab. *10,* 145 (1951).

[180] *Yamatodani, S.,* and *Abe, M.,* Bull. Agric. Chem. Soc. [Japan] *19,* 94 (1955).

[181] Ibid., Bull. Agric. Chem. Soc. [Japan] *21,* 200 (1957).

[182] *Abe, M.,* J. Agric. Chem. Soc. [Japan] *28,* 44 (1954).

[183] *Foster, G. E.,* J. Pharmacy Pharmacol. *7,* 1 (1955).

[184] *Gyenes, I.,* and *Bayer, J.,* Pharmazie *16,* 211 (1961).

[185] *Hampshire, C. H.,* and *Page, G. R.,* Quart. J. Pharm. Pharmacol. *9,* 60 (1936).

[186] *Foster, G. E., MacDonald, J.,* and *Jones, T. S. G.,* J. Pharm. Pharmacol. *1,* 802 (1949).

[187] *Keller, C. C.,* Schweiz. Wschr. Chem. Pharm. *34,* 65 (1896).

[188] *Hofmann, A.,* Helv. Chim. Acta *37,* 314 (1954).

[189] *Rieder, H.P.,* and *Bohmer, M.,* Experientia [Basel] *14,* 463 (1958).

[190] Ibid., Helv. Chim. Acta *42,* 1793 (1959).

[191] *van Urk, H. W.,* Pharm. Week. *66,* 473 (1929).

[192] *Smith, M. I.,* U.S. Publ. Hlth. Rep. *45,* 1466 (1930).

[193] *Schlemmer, F., Wirth, P. H. A.,* and *Peters, H.,* Arch. Pharm. *274,* 16 (1936).

[194] *Voigt, R.,* Mikrochim. Acta [Wien] *1959,* 619.

[195] *Graf, E.,* and *Neuhoff, E.,* Arzneimittel-Forsch. *4,* 397 (1954).

[196] *Rieder, H.P.,* and *Böhmer, M.,* Helv. Chim. Acta *43,* 638 (1960).

[197] *v. Békésy, N.,* Biochem. Z. *302,* 187 (1939).

[198] *Rumpel, W.,* Pharmaz. *10,* 204 (1955).

[199] *Yamatodani, S.,* Ann. Rep. Takeda Res. Lab. *19,* 8 (1960).

[200] *Pöhm, M.,* Arch. Pharm., *286,* 509 (1953).

[201] *Wokes, F.,* and *Crocker, H.,* Quart. J. Pharm. Pharmacol. *4,* 420 (1931).

[202] *Kapoor, A. L., Schumacher, H.,* and *Büchi, J.,* Pharmac. Acta Helvetiae 32, 411 (1957).

[203] Unveröffentlichte Untersuchungen aus den pharmazeutisch-chemischen Forschungs- laboratorien (Unpublished investigations from the pharmaceutical-chemical research laboratories) Sandoz A.G., Basel.

[204] *Stoll, A.,* and *Schlientz, W.,* Helv. Chim. Acta *38,* 585 (1955),

[205] *Bowman, R. L., Caulfield, P.A.,* and *Udenfriend, S.,* Science *122,* 32 (1955).

[206] *Kolšek, J.,* Mikrochim. Acta [Wien] *1956,* 1500.

[207] Ibid. Mikrochim. Acta [Wien] *1956,* 1662.

[208] *Axelrod, J., Brady, R. O., Witkop, B.,* and *Evarts, E. V.,* Ann. New York Acad. Sci. *66,* 435 (1956/1957).

[209] *Hampshire, C. H.,* and *Page, G. R.,* Quart. J. Pharm. Pharmacol. *11,* 57 (1938).

[210] *Schou, S. A.,* and *Bennekou, I.,* Dansk Tidsskr. Farmaci *12,* 257 (1938).

[211] *Schou, S. A.,* and *Jensen, V. G.,* Dansk Tidsskr. Farmaci *22,* 1 (1948).

[212] *Mothes, K., Weygand, F., Gröger, D.,* and *Grisebach, H.,* Z. Naturforsch. *13B,* 41 (1958).

[213] *Silber, A.,* and *Schulze, T.,* Pharmazie *8,* 675 (1953).

[214] *Fuchs, L., Hecht, M.,* and *Pöhm, M.,* Scientia Pharm. *21,* 350 (1953).

[215] *Fischer, R.,* and *Hecht, M.,* Mikrochemie *38,* 538 (1951).

[210] *Hellberg, H.,* Farmac. Revy *50,* 17 (1951).

[217] Ibid. Farmac. Revy *52,* 535 (1953).

[218] *Soffel, W.,* and *Rochelmeyer, H.,* Pharmaz. Ztg. *101,* 1059 (1956).

[219] *van Tamelen, E E.,* Experientia [Basel] *9,* 457 (1953).

[220] *Plieninger, H., Fischer, R., Keilich, G.,* and *Orth, H. D.,* Liebigs Ann. Chem. *642,* 214 (1961).

[221] *Wendler, N. L.,* Experientia [Basel] *10,* 338 (1954).

[222] *Robinson, R.,* The Structural Relations of Natural Products (p. 106), Oxford, Clarendon Press (1955).

[223] *Feldstein, A.,* Experientia [Basel] *12,* 475 (1956).

[224] *Gröger, D., Mothes, K., Simon, H., Floss, H.-G.,* and *Weygand, F.,* Z. Naturforsch. *15B,* 141 (1960).

[225] *Taylor, E. H.,* and *Ramstad, E.,* Nature [London] *188,* 494 (1960).

[226] *Birch, A. J., McLoughlin, B. J.,* and *Smith, H.,* Tetrahedron Letters [London] No. 7, 1 (1960).

[227] *Bhattacharji, S., Birch, A. J., Brack, A., Hofmann, A., Kobel, H., Smith, D. C. C., Smith, H.,* and *Winter, J.,* J. Chem. Soc. [London] *1962,* 421.

[228] *Baxter, R. M., Kandel, S. I.,* and *Okany, A.,* Tetrahedron Letters [London] No. 17, 596 (1961).

[229] *Weygand, F., Floss, H.-G.,* and *Mothes, U.,* Tetrahedron Letters [London] No. 19, 873 (1962).

[230] *Plieninger, H., Fischer, R.,* and *Liede, V.,* Angew. Chem. *74,* 430 (1962).

[231] *Agurell, St.,* and *Ramstad, E.,* Tetrahedron Letters [London] No. 15, 501 (1961).

[232] *Brack, A., Brunner, R.,* and *Kobel, H.,* Arch. Pharmaz. *295,* 510 (1962).

[233] *Stoll, A.,* and *Hofmann, A.,* Helv. Chim. Acta *18,* 421 (1955).

[234] *Semonský, M.,* and *Zikán, V.,* Collect. Czechoslov. Chem. Commun. *25,* 2038 (1960).

[235] *Semonský, M., Černý, A.,* and *Zikán, V.,* Collect. Czechoslov. Chem. Commun. *21,* 382:(1956).

[236] Semonský, M., Zikán, V., and Votava, X., Collect. Czechoslov. Chem. Commun. 22, 1632: (1957).

[237] Zikán, V., Semonský, M., Collect. Czechoslov. Chem. Commun. 24, 12.74 (1959).

[238] Votava, Z., and Semonský, M, and Semonský, M., Arch. int. Pharmacodynam. Thérap. 115, 114 (1958).

[239] Stoll, A., Hofmann, A., Jucker, E., Petrzilka, Th., Rutschmann, J., and Troxler, F., Helv. Chim. Acta 33, 108 (1950).

[240] Stoll, A., and Petrzilka, Th., Helv. Chim. Acta 35, 589 (1952).

[241] Hofmann, A., J. Exp. Med. Sci. [India] 4, 105 (1961).

[242] Garbrecht, W. L., J. Org. Chem. 24, 368 (1959).

[243] Pioch, R. P., U. S. Pat. 2 736 728 (1956).

[244] Paul, R., and Anderson, G. W., J. Amer. Chem. Soc. 82, 4596 (1960).

[245] Černý, A., and Semonský, M., Collect. Czechoslov. Chem. Commun. 27, 1585 (1962).

[246] Schweiz. Pat. Registration, Sandoz A. G., Basel.

[247] Semonský, M., and Zikán, V., Collect. Czechoslov. Chem. Commun. 25, 1190 (1960).

[248] Zikán, V., and Semonský, M., Collect. Czechoslov. Chem. Commun. 25, 1922 (1960).

[249] Belg. Pat. 609 010; French Pat. 1 303 288, Sandoz A.G., Basel.

[250] Belg. Pat. 607 502.; French Pat. 1 298 661, Sandoz A.G., Basel.

[251] Votava, Z., and Lamplova, I., Proceedings of the 2nd Meeting of the Collegium Internationale Neuro-Psychopharmacologicum, Basel (1960). In: Neuro-Psychopharmacology 2, 68 (1961). Ed. by: Rothlin, E., Elsevier-Publ. Co., Amsterdam-London-New York-Princeton.

[252] Votava, Z., and Lamplova, I., Physiol. Bohemosloven. 12, 37 (1963).

[253] Troxler, F., and Hofmann, A., Helv. Chim. Acta 40, 1706 (1957).

[254] Ibid., Helv. Chim. Acta 40, 1721 (1957).

[255] Belg. Pat. 607 294; French Pat. 1 297 632, Sandoz A.G., Basel.

[256] Troxler, F., and Hofmann, A., Helv. Chim. Acta 40, 2160 (1957).

[257] Cerletti, A., and Konzett, H., Arch. Exp. Path. Pharmakol. 228, 146 (1956).

[258] Cerletti, A., and Rothlin, E., Nature [London] 176, 785 (1955).

[259] Zehnder, K., and Cerletti, A., Helv. Physiol. Pharmacol. Acta 14, 264 (1956).

[260] Berde, B., and Cerletti, A., Helv. Physiol. Pharmacol. Acta 14, 325 (1956).

[261] German Patent Nr. 1 015 810 on 5.1.1956, Sandoz A.G., Basel.

[262] Hellberg, H., Acta Chem. Scand. 11, 219 (1957).

[263] Ibid. Pharm. Week. 93, 1 (1958).

[264] Ibid. Acta Chem. Scand. 12, 678 (1958).

[265] Ibid. Acta Chem. Scand. 13, 1106 (1959).

[266] Ibid. Acta Chem. Scand. 16, 1363 (1962).

[267] Kline, G. B., Fornefeld, E. J., Chauvette, R.R., and Kornfeld, E. C., J. Org. Chcm. 25,142 (1960).

[268] *Stadler, P.A., Frey, A. J.,* and *Troxler, F.,* Chimia [Zürich] *15,* 575 (1961).

[269] *Wenzing, M.,* Liebigs Ann. Chem. *239,* 239 (1887).

[270] *Little, J. S., Taylor, W. I.,* and *Thomas, B. R.,* J. Chem. Soc. [London] *1954,* 4036.

[271] French Pat. 1 298 156, Sandoz A.G., Basel.

[272] *Freter, K., Axelrod, J.,* and *Witkop, B.,* J. Amer. Chem. Soc. 79, 3191 (1957).

[273] *Troxler, F.,* and *Hofmann, A.,* Helv. Chim. Acta *42,* 793 (1959).

[274] *Hofmann, A., Heim, R., Brack, A., Kobel, H., Frey, A., Ott, H., Petrzilka, Th.,* and *Troxler, F.,* Helv. Chim. Acta *42,* 1557 (1959).

[275] *Brack, A., Hofmann, A., Kalberer, F., Kobel, H.,* and *Rutschmann, J.,* Arch. Pharmaz. *294/66.* Bd., 230 (1961).

[276] *Brack, A., Brunner, R.,* and *Kobel, H.,* Helv. Chim. Acta *45,* 276 (1962).

[277] *Axelrod, J., Brady, R. O., Witkop, R.,* and *Evarts, E. V.,* Nature [London] *178,* 143 (1956).

[278] *Rothlin, E.,* The Pharmacology of the Natural and the Dihydrogenated Alkaloids of Ergot, Bull. Schwciz. Akad. Med. Wiss. 2, 249 (1946/1947).

[279] *Lanz, U., Cerletti, A.,* and *Rothlin, E.,* Helv. Physiol. Pharmacol. Acta *13,* 207 (1955).

[280] *Stoll, A., Rutschmann, J.,* and *Hofmann, A.,* Helv. Chim. Acta *37,* 820 (1954).

[281] *Stoll, A., Rothlin, E., Rutschmann, J.,* and *Schalch, W.R.,* Experientia [Basel] *11,* 396 (1955).

[282] *Boyd, E. S.,* Arch. Int. Pharmacodynam. *120,* 292 (1959).

[283] *Slaytor, M. B.,* and *Wright, S. E.,* J. Med. Pharm. Chem. *5,* 483 (1962).

[284] *Brown, R. D.,* and *Coller, B. A. W.,* Austr. J. Chem. *12,* 152 (1959).

[285] *Settimj, G., Weber, H.,* and *Arigoni, D.,* Helv. Chim. Acta (in Print).

[286] *Curry, A. S.,* J. Pharm. Pharmacol. *11,* 411 (1959).

[287] *Voigt, R.,* and *Kaehler, A.,* Pharmaz. Zentralhalle *101,* 95 (1962).

[288] *Taber, W. A.,* and *Vining, L. C.,* Chem. and Ind. *1959,* 1218.

[289] *Gröger, D., Wendt, H. J., Mothes, K.,* and *Weygand, F.,* Z. Naturforsch. *14B,* 355 (1959).

[290] *Stoll, A.,* and *Rüegger, A.,* Helv. Chim. Acta *37,* 1725 (1954).

[291] *Klavehn, M.,* and *Rochelmeyer, H.,* Dtsch. Apotheker-Ztg. *101,* 477 (1961).

[292] *Schindler, R.,* and *Bürgin, A.,* Helv. Chim. Acta *39,* 2132 (1956).

[293] *Harley-Mason, J.,* Chem. and Ind. *1954,* 251.

[294] *Allport, N. L.,* and *Jones, N. R.,* Quart. J. Pharm. Pharmacol. *14,* 106 (1941).

[295] *Gröger, D., Mothes, K., Simon, H., Floss, H. G.,* and *Weygand, F.,* Z. Naturforsch. *16B,* 432 (1961).

[296] *Gröger, D.,* Arch. Pharm. *292,* 649 (1959).

[297] *Baxter, R. M., Kandel, S. I.,* and *Okany, A.,* Chem. and Ind. *1961,* 1453.

[298] *Doepfner, W.,* Experientia [Basel] *18,* 256 (1962).

[299] *Leemann, H. G., Stich, K.,* and *Thomas, M.,* Physicochemical methods in pharmaceutical chemistry. I. Spektrofluorometry. Fortschr. Arzneimittelforsch. 6, 151 (1963), Birkhauser-Verlag, Basel.

[300] *Malin, B.*, and *Westhead, I.*, J. Biochem. Microbiol. Technol. Engng. *1*, 49 (1959).

[301] *Hofmann, A.*, Unveröffentlichte Untersuchungen aus den pharmazeutisch-chemischen Forschungslaboratorien (Unpublished studies from the pharmaceutical-chemical research laboratories) Sandoz A.G., Basel.

[302] *Pöhm, M.*, Arch. Pharm., *291*,468 (1958).

[303] *Guggisberg, H.*, Schweiz. Med. Wschr. *51*, 750 (1921).

[304] Ibid. Schweiz. Med. Wschr. *54*, 97 (1924).

[305] *Rothlin, E.*, Arch. Int. Pharmacodyn. 27, 459 (1923).

[306] Ibid. Klin. Wschr. *4*, 1437 (1925).

[307] Ibid. Bull. Gén. Thér. Paris, *176*, 416 (1925).

[308] Ibid. J. Pharm. Exp. Ther. *36*, 657 (1929).

[309] *Rothlin, E.*, and *Cerletti, A.*, Helv. Physiol. Pharmacol. Acta 7, 333 (1949).

[310] Ubersicht and Zugang zu den Original-Publikationen (For an overview and access to the original publications, see:) *Rothlin, E.*, and *Fanchamps, A.*, Revue de Pathologie Génétale et Comparée, *55*, 1427 (1955).

[311] *Goodmann, L. S.*, and *Gilman, A.*, The Pharmacological Basis of Therapeutics. 2nd Ed., The Macmillan Co., New York (1955), p. 577ff., p. 886 ff.

[312] *Cerletti, A.*, Proceedings of the 1st International Congress of Neuro-Pharmacology, Rome (1958). In: Neuro-Psychopharmacology. Ed. by: *Bradley, P. B., Deniker, P.*, and *Radouco-Thomas, C.*, Elsevier Publ. Co., Amsterdam-London-New York-Princeton (1959), p. 117.

[313] *Rothlin, E.*, Schweiz. Med. Wschr. *52*, 978 (1922).

[314] *Kopet, J.C.*, and *Dille, J.M.*, J. Amer. Pharm. Assoc. Sci. *31*, 109 (1942).

[315] *Rothlin, E.*, Helv. Chim. Acta *29*, 1290 (1946).

[316] *Konzett, H.*, and *Rothlin, E.*, Brit. J. Pharmacol. Chemotherapy 8, 201 (1953).

[317] *Taeschler, M., Cerletti, A.*, and *Rothlin, E.*, Helv. Physiol. Pharmacol. Acta *10*, 120 (1952).

[318] *Hofmann, A.*, Die Geschichte des LSD 25 (The history of LSD 25). Triangel, Sandoz-Zeitschrift fur Medizinische Wissenschaften 2, 117 (1955).

[319] Ibid. Acta Physiol. Pharmacol. Neerl. 8, 240 (1959).

[320] *Stoll, W. A.*, Schweiz. Arch. Neurol. Psychiat. *60*, 279 (1947).

[321] *Späth, E.*, Mh. Chem. *40*, 129 (1919).

[322] *Hofmann, A.*, Chimia [Zürich] *14*, 309 (1960).

[323] *Solms, H.*, Praxis [Bern] *45*, 746 (1956).

[324] Ibid. J. Clin. Exp. Psychopath. and Quart. Rev. Psychiat. *17*, 429 (1956).

[325] *Cerletti, A.*, Helv. Med. Acta *25*, 330 (1958).

[326] Ibid. Über Vorkommen und Bedeutung der Indolstruktur in der Medizin and Biologie (Occurrence and importance of the indole structure in medicine and biology). Fortschr. Arzneimittel-Forsch. 2, 227 (1960).

[327] *Rothlin, E., Cerletti, A., Konzett, H., Schalch, W. R.*, and *Taeschler, M.*, Experientia [Basel] *12*, 154 (1956).

[328] *Rothlin, E.*, J. Pharm. Pharmacol. [London] 9, 569 (1957).

[329] *Neuhold, K., Taeschler, M.,* and *Cerletti, A.,* Helv. Physiol. Pharmacol. Acta *15,* 1 (1957).

[330] *Rothlin, E.,* and *Cerletti, A.,* in *L. Cholden:* Lysergic Acid Diethylamide and Mescaline in Experimental Psychiatry, S. 1–7, Grune and Stratton, New York-London (1956).

[331] *Haley, T. J.,* and *Rutschmann, J.,* Experientia [Basel] *13,* 199 (1957).

[332] *Busch, A. K.,* and *Johnson, W. C.,* Dis. Nerv. Syst. *11,* 241 (1950).

[333] *Sandison, R. A.,* J. Mental Sci. *100,* 508 (1954).

[334] *Frederking, W.,* Psyche [Heidelberg] 7, 342 (1953/1954).

[335] *Leuner, H.,* and *Holfeld, H.,* Psychiat. Neurol. [Basel] *143,* 379 (1962).

[336] *Isbell, H., Miner, H.J.,* and *Logan, C. R.,* Psychopharmacologia [Berlin] *1,* 20 (1959).

[337] *Hofmann, A.,* Relationship Between Spatial Arrangement and Mental Effects, in *Rinkel, M.,* and *Denber, H. C. B.:* Chemical Concepts of Psychosis. McDowell-Obolensky, New York (1958), p. 85.

[338] *Rothlin, E.,* Ann. New York Acad. Sci. *66,* 668 (1957).

[339] Unveröffentlichte Untersuchungen aus den pharmakologischen Forschungslaboratorien (Unpublished studies from the pharmacological research laboratories), Sandoz A.G., Basel.

[340] *Cerletti, A.,* and *Doepfner, W.,* J. Pharm. Exp. Ther. *122,* 124 (1958).

[341] *Doepfner, W.,* and *Cerletti, A.,* Int. Arch. Allergy, N. Y. *12,* 89 (1958).

[342] *Cerletti, A.,* and *Doepfner, W.,* Helv. Physiol. Pharmacol. Acta *16,* C 55–C 57 (1958).

[343] *Fanchamps, A., Doepfner, W., Weidmann, H.,* and *Cerletti, A.,* Schweiz. Med. Wschr. *90,* 1040 (1960).

[344] *Lanz, R.,* Schweiz. Med. Wschr. *90,* 1046 (1960).

[345] *Friedmann, A. P.,* Ann. New York Acad. Sci. *86,* 216 (1960).

[346] *Heyck, H.,* Schweiz. Med. Wschr. *90,* 203 (1960).

[347] *Sicuteri, F.,* Int. Arch. Allergy *15,* 300 (1959).

[348] *Abbott, K. H.,* Dis. Nerv. Syst. *23,* 579 (1962).

[349] *Abramson, H. A.,* J. Psychol. *41,* 199 (1956).

[350] *Spiro, K.,* Schweiz. Med. Wschr. *51,* 737 (1921).

[351] *Brügger, J.,* Helv. Physiol. Pharmacol. Acta *.3,* 117 (1945).

[352] Wolff, *H. G.,* Headache and Other Head Pain. Oxford University Press, New York, 1948.

[353] *Rothlin, E.,* and *Cerletti, A.,* Verh. Dtsch. Ges. Kreislaufforsch. *15,* 158 (1949).

[354] *Rothlin, E.,* Wien. Klin. Wschr. *62,* 893 (1950).

[355] Ibid. Rev. Pharmacol. Ther. Exp. *1,* 212 (1929),

[356] *Gaddum, J. H.,* J. Physiol. *121,* 15 P (1953).

[357] *Fingl, E.,* and *Gaddum, J. H.,* Fed. Proc. *12,* 320 (1953).

[358] *Baxter, R. M., Kandel, S. I., Okany, A.,* and *Tam, K. L.,* J. Amer. Chem. Soc. *84,* 4350 (1962).

[359] *Gröger, D.,* Arch. Pharmaz., *292,* 389 (1959).

[360] Ibid. Pharmazie *15*, 715 (1960).

[361] *Brady, L. R.,* and *Tyler Jr., V. E.,* J. Amer. Pharmac. Assoc. *49*, 332 (1960).

[362] *Gröger, D.,* and *Erge, D.,* Planta Med. [Stuttgart] *9*, 471 (1961).

[363] *Agurell, St.,* and *Ramstad, E.,* Arch. Biochem. Biophysics *98*, 457 (1962).

[364] *Mothes, K., Winkler, K., Gröger, D., Floss, H. G., Mothes, U.,* and *Weygand, F.,* Tetrahedron Letters [London] *1962*, 933.

[365] *Pacifici, L. R., Kelleher, W. J.,* and *Schwarting, A. E.,* Lloydia *25*, 37 (1962).

[366] *Teuscher, E.,* Pharmazie *16*, 570 (1961).

[367] *Arcamone, F., Chain, E. B., Ferretti, A., Minghetti, A., Pennella, P.,* and *Tonolo, A.,* Biochim. Biophys. Acta *57*, 174 (1962).

[368] *Glasser, A.,* Nature [London] *189*, 313 (1961).

[369] *Grob, G. A.,* and *Meier, W.,* Helv. Chim. Acta *39*, 776 (1956).

[370] *Plieninger, H.,* and *Suhr, K.,* Chem. Ber. *90*, 1980 (1957).

[371] *Plieninger, H.,* and *Müller, W.,* Chem. Ber. *93*, 2024 (1960).

[372] Ibid., Chem. Ber. *93*, 2029 (1960).

[373] *Uhle, F. C., Vernick, C. G.,* and *Schmir, G. L.,* J. Am. Chem. Soc. *77*, 3334 (1955).

[374] *Uhle, F. C.,* and *Robinson, S. H.,* J. Am. Chem. Soc. *77*, 3544 (1955).

[375] *Uhle, F. C.,* and *Harris, L. S.,* J. Am. Chem. Soc. *79*, 102 (1957).

[376] *Mann, F. G.,* and *Tetlow, A. J.,* Chem. and Ind. *1953*, 823.

[377] *Pöhm, M.,* Arch. Pharmaz. *289*, 324 (1956).

[378] *Alexander, T. G.,* and *Banes, D.,* J. Pharm. Sci. *50*, 201 (1961).

[379] *Macek, K.,* and *Vanecek, S.,* Collect. Czechoslov. Chem. Commun. *24*, 315 (1959).

[380] *Agurell, St.,* and *Ramstad, E.,* Lloydia *25*, 67 (1962).

[381] *Rochelmeyer, H., Stahl, E.,* and *Patani, A.,* Arch. Pharmaz. *291*, 1 (1958).

[382] *Voigt, R.,* Pharmazie *13*, 294 (1958).

[383] *Taylor, E. H.,* and *Ramstad, E.,* J. Pharm. Sci. *50*, 681 (1961).

[384] *Baxter, R. M., Kandel, S. I.,* and *Okany, A.,* J. Am. Chem. Soc. *84*, 2997 (1962).

[385] U.S.Pat. 3 085 092, Sandoz A.G., Basel.

[386] *Voigt, R.,* Scientia Pharmac. *28*, 123 (1960).

[387] Ibid. and *Wichmann, D.,* Pharmazie *16*, 35, 369 (1961).

[388] Ibid. Pharmazie *17*, 101,156 (1962).

[389] *Wichmann, D.,* and *Voigt, R.,* Pharmazie *17*, 411 (1962).

[390] *Winkler, K.,* and *Gröger, D.,* Pharmazie *17*, 658 (1962).

[391] *Stahl, E.,* Dünnschicht-Chromatographie, Springer-Verlag, Berlin, 1962.

[392] Eine zusammenfassende Darstellung der Biochemie, Physiologie und Pharmakologie des im Warmblüterorganismus weitverbreiteten, insbesonders auch in gewissen Partien des Gehirns angereicherten Serotonins (5-Hydroxytryptamin) findet man bei *Cerletti, A.,* Helv. Med. Acta *25*, 330 (1958). (A comprehensive description of the biochemistry, physiology and pharmacology of serotonin (5-hydroxytryptamine), which is widespread in warm-

blooded organisms and is particularly concentrated in certain parts of the brain, can be found in Cerletti, A., Helv. Med. Acta 25, 330 (1958).)

[393] *Stadler, P. A., Frey, A. J.,* and *Hofmann, A.,* Helv. Chim. Acta 46, 2300 (1963).

[394] Swiss Patent Registration, Sandoz A.G., Basel.

AUTHOR INDEX

Numbers in parentheses refer to the corresponding citations in the Bibliography on page 223.

Baxter, R. M., Kandel, S. I., Okany, A., and Tam, K. L.: 134 (358)

Bayer, J.: see Gyenes, I, 113, 117 (184)

Becker, B.: see Stoll, A., 74 (160)

Becker, B.: see Stoll, A., 77 (163)

v. Békésy, N.: 7 (37)

v. Békésy, N.: 116 (197)

v. Békésy, N.: 7 (40)

v. Békésy, N.: 5 (22)

Benedikt VIII: 9

Bennekou, I.: see Schou, S. A., 127 (210)

Beran, M.: see Semonský, M., 101 (178)

Berde, B., and Cerletti, A.: 160, 193 (260)

Berde, B.: see Schlientz, W., 21, 22, 129 (103)

Bergel, F.: see Atherton, F. R., 60, 71 (139)

Bhattacharji, S., Birch, A. J., Brack, A., Hofmann, A., Kobel, H., Smith, D. C. C., Smith, H., and Winter, J.: 132 (227)

Birch, A. J., McLaughlin, B. J., and Smith, H.: 132 (226)

Birch, A. J.: see Bhattacharji, S., 132 (227)

Bischoff, W.: see Silber, A., 7 (43)

Bischoff, W.: see Silber, A., 4 (16)

Böhmer, M.: see Rieder, H.P., 114 (189)

Böhmer, M.: see Rieder, H.P., 114 (190)

Böhmer, M.: see Rieder, H.P., 116 (196)

Bonino, C.: see Arcamone, F., 5, 6, 33, 34, 35 (20)

Bowman, R. L., Caulfield, P. A., and Udenfriend, S.: 125 (205)

Boyd, E. S.: 174, 188 (282)

Brack, A.: see Stoll, A., 7 (38)

Brack, A.: see Stoll, A., 7 (39)

Brack, A.: see Stoll, A., 5, 6, 15, 96, 101 (17)

Brack, A.: see Hofmann, A., 5, 6, 15, 99, 100, 101, 105, 110, 171 (18)

Brack, A.: see Hofmann, A., 33, 171 (274)

Brack, A., Hofmann, A., Kalberer, F., Kobel, H., and Rutschmann, J.: 171 (275)

Brack, A., Brunner, R., and Kobel, H.: 134 (232)

Brack, A., Brunner, R., and Kobel, H.: 171 (276)

Brack, A.: see Kobel, H., 6, 105 (35)

Brack, A.: see Bhattacharji, S., 132 (227)

Brady, R. O.: Axelrod, J., 173 (277)

Brady, R. O.: see Axelrod, J., 126,173 (208)

Brady, L. R., and Tyler, jr., V. E.: 6 (361)

Brady, L. R.: see Abou-Chaar, C. I., 35 (115)

Brady, L. R.: 4 (15)

Brejcha, V.: see Kybal, J., 5 (21)

Brown, R. D. and Coller, B. A. W.: 174 (284)

Brügger, J.: 180 (351)

Brunner, R.: see Stoll, A., 5, 6, 15, 96,101 (17)

Brunner, R.: see Hofmann, A., 5, 6, 15, 99, 100,101,105,110,171 (18)

Brunner, R.: see Schlientz, W., 21, 22, 129 (103)

Brunner, R.: see Schlientz, W., 21, 22 (104)

Brunner, R.: see Kobel, H., 6,105 (35)

Brunner, R.: see Brack, A., 134 (232)

Brunner, R.: see Brack, A., 171 (276)

Brunner, R.: see Schlientz, W., 15, 28, 29, 74, 75, 92, 93 (80)

Büchel, K. G.: see Rassbach, H., 6 (31)

Büchi, J.: see Kapoor, A. L., 117, 128 (202)

Burckhardt, E.: see Stoll, A., 14, 31 (65)

Burckhardt, E.: see Stoll, A., 14, 31 (66)

Burckhardt, E.: see Stoll, A., 15, 23, 24 (70)

Bürgin, A.: see Schindler, R., 129 (292)

Busch, A. K., and Johnson, W. C.: 189 (332)

Cahn, R. S., Ingold, C. K., and Prelog, V.: 58 (138)

Campbell, W. P.: 5 (23)

Carr, F. H.: see Barger, G., 14, 75 (63)

Carr, F. H.: see Barger, G., 12, 14, 23 (52)

Carranza, J.M.: see Lindquist, J.C., 4 (14)

Caulfield, P.A.: see Bowman, R. L., 125 (205)

Cerletti, A.: see Rothlin, E., 176, 178 (309)

Cerletti, A.: see Rothlin, E., 182 (353)

Cerletti, A.: see Taeschlcr, M., 183 (317)

Jacobs, W. A., and Craig, L. C.: 15, 35, 39 (83)

Jacobs, W. A., and Craig, L. C.: 16, 72 (84)

Jacobs, W. A., and Craig, L. C.: 16 (85)

Jacobs, W. A., and Craig, L. C.: 16, 74, 76 (86)

Jacobs, W. A., and Craig, L. C.: 16, 74 (87)

Jacobs, W. A., and Craig, L. C.: 39 (124)

Jacobs, W. A., Craig, L. C., and Rothen, A.: 40 (126)

Jacobs, W. A., and Craig, L. C.: 74 (159)

Jacobs, W. A., and Craig, L. C.: 39 (123)

Jacobs, W. A., and Craig, L. C.: 38 (119)

Jacobs, W. A., and Gould jr., R. G.: 16, 40, 58 (101)

Jacobs, W. A., and Craig, L. C.: 16, 40 (90)

Jacobs, W. A., and Craig, L. C.: 76 (161)

Jacobs, W. A.: see Craig, L. C., 40 (127)

Jacobs, W. A., and Gould jr., R. G.: 40 (128)

Jacobs, W. A., and Craig, L. C.: 39 (125)

Jacobs, W. A., and Gould jr., R. G.: 40 (129)

Jacobs, W. A.: see Gould jr., R. G., 60 (143)

Jacobs, W. A.: see Gould jr., R. G., 40 (130)

Jacobs, W. A.: see Uhle, F. C., 16, 59, 65 (91)

Jensen, V. G.: see Schou, S. A., 127 (211)

Johnson, W. C.: see Busch, A. K., 189 (332)

Jones, N. R.: see Allport, N. L., 127 (294)

Jones, R. G.: see Kornfeld, E. C., 16, 68 (95)

Jones, R. G.: see Kornfeld, E. C., 16, 36, 68, 166, 169 (96)

Jones, T. S. G.: see Foster, G. E., 113, 128 (186)

Jucker, E.: see Stoll, A., 140 (239)

Kaehler, A.: see Voigt, R., 128 (287)

Kaiser, H. and Thielmann, E.: 6 (30)

Kalberer, F.: see Brack, A., 171 (275)

Kandel, S. I.: see Baxter, R. M., 132 (228)

Kandel, S. I.: see Baxter, R. M., 131 (297)

Kandel, S. I.: see Baxter, R. M., 132 (384)

Kandel, S. I.: see Baxter, R. M., 134 (358)

Kapoor, A. L., Schumacher, H., and Büchi, J.: 117, 128 (202)

Kawatani, T.: 4 (11)

Keilich, G.: see Plieninger, H., 131, 134 (220)

Kelleher, W. J.: see Pacifici, L. R., 6 (365)

Keller, C. C.: 114 (187)

Kharasch, M. S., and Legault, R. R.: 14, 31 (67)

Klavehn, M., and Rochelmeyer, H.: 130 (291)

Kline, G. B.: see Kornfeld, E. C., 16, 68 (95)

Kline, G. B.: see Kornfeld, E. C., 16, 36. 68, 166, 169 (96)

Kline, G. B., Fornefeld, E. J., Chauvette, R.R., and Kornfeld, E. C.: 166 (267)

Kobel, H.: see Stoll, A., 5, 6, 15, 96, 101 (17)

Kobel, H.: see Hofmann, A., 5, 6, 15, 99,100, 101, 105, 110, 171 (18)

Kobel, H.: see Hofmann, A., 33, 171 (274)

Kobel, H.: see Brack, A., 171 (275)

Kobel, H.: see Brack, A., 171 (276)

Kobel, H., Brunner, R., and Brack, A.: 6, 105 (35)

Kobel, H.: see Bhattacharji, S., 132 (227)

Kobel, H.: see Brack, A., 134 (232)

Kobert, R.: 8 (44)

Kolsek, J.: 129 (206)

Kolsek, J.: 129 (207)

Komatsu, H.: see Abe, M., 15, 29, 30 (75)

Konzett, H., and Rothlin, E.: 183 (316)

Konzett, H.: see Cerletti, A., 160, 193 (257)

Konzett, H.: see Rothlin, E., 187 (327)

Kopet, J.C., and Dille, J.M.: 181 (314)

Kornfeld, E. C., Fornefeld, E. J., Kline, G.B., Mann, M. J., Jones, R. G., and Woodward, R. B.: 16, 68 (95)

Kornfeld, E. C., Fornefeld, E. J., Kline, G. B., Mann, M. J., Morrison, D. E., Jones, R. G., and Woodward, R. B.: 16, 36, 68, 166, 169 (96)

Smith, S., and Timmis, G. M.: 32 (110)

Smith, S., and Timmis, G. M.: 32 (109)

Smith, S., and Timmis, G. M.: 16, 33, 35, 36, 38 (88)

Smith, S.: see Grant, R. L., 31 (107)

Smith, S., and Timmis, G. M.: 14, 22 (69)

Soffel, W., and Rochelmeyer, H.: 128 (218)

Solms, H.: 185 (323)

Solms, H.: 185 (324)

Späth, E.: 184 (321)

Spilsbury, J. F., and Wilkinson, S.: 15, 33, 95, 102, 104, 111 (79)

Spiro, K., and Stoll, A.: 12, 176, 178 (62)

Spiro, K.: 176 (350)

Stadler, P.A., Frey, A. J., and Troxler, F.: 166, 169 (268)

Stadler, P. A., and Hofmann, A.: 16, 57 (99)

Stadler, P. A., Frey, A. J., and Hofmann, A.: 87 (393)

Stadler, P.A.: see Hofmann, A., 87 (171)

Stadler, P. A.: see Schlicntz, W., 15, 28, 29, 74, 75 (80)

Stahl, E.: see Rochelmeyer, H., 129 (381)

Stahl, E.: 129 (391)

Stearns, J.: 11, 176 (48)

Stenlake, J. B.: 47, 50 (134)

Stich, K.: see Leemann, H. G., 126 (299)

Stoll, A.: 12, 14 (56)

Stoll, A.: see Spiro, K., 12, 176, 178 (62)

Stoll, A.: 12, 14, 176 (58)

Stoll, A.: 12, 14, 176 (59)

Stoll, A.: 12, 14, 176 (60)

Stoll, A., and Burckhardt, E.: 14, 31 (66)

Stoll, A., and Burckhardt, E.: 14, 31 (65)

Stoll, A., and Burckhardt, E.: 15, 23, 24 (70)

Stoll, A,. and Hofmann, A.: 36, 44, 72 (116)

Stoll, A., and Hofmann, A.: 16, 72 (89)

Stoll, A., and Hofmann, A.: 36, 44, 70, 72 (117)

Stoll, A., and Hofmann, A.: 70, 72, 74, 135, 137, 148, 181, 184 (157)

Stoll, A., and Hofmann, A.: 15, 23, 24, 25, 26, 27, 75 (71)

Stoll, A., Hofmann, A., and Becker, B.: 74 (160)

Stoll, A., and Hofmann, A.: 38, 44, 125, 161, 182 (120)

Stoll, A., and Brack, A.: 7 (38)

Stoll, A., and Brack, A.: 7 (39)

Stoll, A.: 12, 14, 18, 19, 21, 74 (61)

Stoll, A.: Hofmann, A., and Petrzilka, Th.: 38, 39, 44, 62, 63, 137, 148, 161 (121)

Stoll, A., Hofmann, A., and Troxler, F.: 16, 41, 58, 109, 112 (92)

Stoll, A., Hofmann, A., and Schlientz, W.: 98 (176)

Stoll, A., Petrzilka, Th., and Becker, R.: 77 (163)

Stoll, A. and Rutschmann, J.: 60 (142)

Stoll, A., Hofmann, A., Juckcr, E., Petrzilka, Th., Rutschmann, J., and Troxler, F.: 140 (239)

Stoll, A., Rutschmann, J., and Schlientz, W.: 16, 66, 67 (93)

Stoll, A., and Hofmann, A.: 33, 35, 78 (111)

Stoll, A., Pctrzilka, Th., and Rutschmann, J.: 71 (158)

Stoll, A., Hofmann, A., and Petrzilka, Th.: 16, 79, 81 (94)

Stoll, A.: III (1)

Stoll, A.: III (2)

Stoll, A., and Petrzilka, Th.: 140 (240)

Stoll, A., Petrzilka, Th., and Rutschmann, J.: 68 (156)

Stoll, A., and Petrzilka, Th.: 64, 71 (145)

Stoll, A., and Petrzilka, Th.: 64 (146)

Stoll, A., and Rutschmann, J.: 63, 66 (144)

Stoll, A., and Rutschmann, J.: 66 (147)

Stoll, A., Rutschmann, J., and Hofmann, A.: 174, 188 (280)

Stoll, A., and Rüegger, A.: 128 (290)

Stoll, A., Brack, A., Kobel, H., Hofmann, A., and Brunner, R.: 5, 6, 15, 96, 101 (17)

Stoll, A., Petrzilka, Th., Rutschmann, J., Hofmann, A., and Günthard, Hs. H.: 16, 47, 50 (97)

Stoll, A., Rothlin, E., Rutschmann, J., and Schalch, W. R.: 174, 188 (281)

CHEMICAL INDEX

GENERAL INDEX

Absolute configuration, 20, 56, 63, 65, 67, 96, 99-101, 120, 176
  at C12', 101
  at C2', 99
Absorption, 49-51, 126, 128, 132-134, 141-143, 176, 194
Acetylcholine, 14, 212
Acid amine linkage, 78
Acidity, 60, 102
Acta Sanctorum, 10
Active ingredient content determination, 189
Adrenaline antagonism, 210
Adrenergic system, 192
Adrenolytic, 26, 190, 195, 198, 209-210, 213
African foxtail millet (*Pennisetum typhoideum* Rich.), 5-6, 18, 107-108, 112-114
Agropyrum, 6, 18, 106-107, 110, 112-113, 115, 117-118
  Agropyrum ciliare Fr., 107
  Agropyrum semicostatum Nees, 18, 107, 113
Alkaline hydrolysis, 80, 126
Alkaloid, 5-8, 13-14, 17-20, 22-24, 34-37, 42, 83, 86-87, 90-91, 94, 99, 101, 103-105, 108-109, 111-112, 115-118, 126-128, 131, 144, 148, 153, 184, 189, 193, 195-196
  preparation, 13, 17
  production, 7
Alkanolamide type, 21
Alkylation, 169, 208
Allergy, 214
Aluminum oxide column, 125, 144
Aluminum sulfate, 23
Amide type, 21
Ampoule, 196
Analeptic, 198

Animal, 143, 160, 183, 185-186, 193, 205
  experiments, 205
  organism, 183, 185-186
Antagonism, 170, 186, 190, 193, 201-202, 212
  to serotonin, 201-202
  to sympathetically inhibited function 193
Antihypertensive, 191, 210
Antipsychotic, 204
Antipyretic, 198
Antiserotonin effect, 211, 213
Antonites, 10
Aquitaine, 10
Artificial, 6-8, 15, 23, 198
  cultivation, 6-7, 218
  infection, 7, 24
  inoculation, 8, 15
  parasitic cultivation, 7
Ascospores, 1, 4
Ascus (pl. asci), 1
Aspergillus, 19, 40, 107, 115, 117
  Aspergillus fumigatus Fres., 115, 117
Asymmetry, 48-52, 56, 63-64, 72, 81, 85, 93, 96, 101, 120, 123, 179
  Assymetric carbon, 49-50
  Center of asymmetry, 48-50, 54, 63-64, 81, 93, 179
Austria, 8
Autonomic nervous system, 14, 189, 194, 200
Axial arrangement, 57, 63
Aztec, 40, 199

Badoh negro, 39
Badoh, 39
Balkan, 6
Barbiturate, 202
Basel, 8, 12, 23, 36, 202

Reductive, 77, 90-91, 93, 123
Thermal, 93, 102
Clinical, 156, 160, 189, 195, 197, 199,
214
application, 156, 197
trials / human clinical trials, 160
Color reaction and colorimetric
determination, 24, 26-32, 34-35, 48,
72, 81, 108, 110, 129-133, 145, 172,
176, 178, 195
Configuration, 20, 53, 56, 58, 63-65, 67,
85, 93, 99-102, 105, 120, 125, 127
Conformation, 62, 123
Conidia, 3-4, 7
Conjugation, 47-48, 50
Consciousness, 200
Repressed contents of consciousness,
204
Constantinople, 10
Contamination, 5, 134
Contractile effect, 14
Uterine contraction, 190, 201
Vasoconstriction, 190, 197, 201
Convolvulaceae, 39-40, 42, 107
Convulsions, 10, 189
Convulsive ergotism (ergotism
convulsivus), 9, 10
Cosmic wealth of images, 201
Crystallization, 24, 29, 32, 37, 42-44, 46,
84, 98, 159-162, 164
equilibrium, 24, 29, 32, 42-44, 98,
159
Crystallized alkaloid preparation, 17
Fractional crystallization, 32-33, 43,
84, 98, 100
Cultivation, 6-7
Artificial parasitic cultivation, 7
Cultivation in vitro, 6-7, 15, 193, 212
Cultivation on artificial nutrient
media, 6
Culture, 5-7, 149, 151-152, 184
Culture media, 6
Saprophytic culture, 5-7, 149, 151
Submerged culture, 149
Surface culture, 6
Curtius azide method, 83

Cyclol, 20, 94-98, 101, 103, 159
Cyperaceae, 30

Death, 189, 203
Decomposability, 23, 45, 51, 128
Dehydrogenation, 80, 181
of indoline into indole derivative, 48,
68, 71, 182
with Raney nickel, 80
Delysid, 156, 198
Demethylation of lysergic acid, 67
Depressive effects in animals, 202
Des-base, 51
Deseril, 170, 212, 214
Deuterium, 148
Dextrorotation, 22, 33, 42-43, 100, 144,
175
Dextrorotatory ergot alkaloids and
lysergic acid derivatives, 175-176
Specific dextrorotation, 42, 49, 51, 78
Di-(p-toluyl)-tartaric acid, 84
Dialkylamide, 62
Diastereomer, 100, 125
Diastereomeric mixture, 52, 123
Diencephalon, 191
Dimorphism, 37
Dioxindole derivatives, 182
Distillate, 94
Distribution, 143-144, 186, 203
Distribution in animal organism, 183,
185-186
Distribution in animal tissue, 143
Dysmenorrhea, 197

Edema test, 212, 214
Effective dose, 191-192, 203
Electroencephalogram (EEG), 202
Electrostatic effect, 60
Elymus, 5-6, 18, 30, 106, 108-110, 112,
118
Elymus canadensis, 5
Elymus mollis Tri, 18, 30, 106, 108-
109, 112
Emetic, 194, 210
Emission spectrum, 141-143
England, 5, 205